Books are to be returned on or before
the last date below.

D1759261

LIBREX–

Steve Jones
General Editor

Vol. 9

Peter Lang
New York • Washington, D.C./Baltimore • Bern
Frankfurt Am Main • Berlin • Brussels • Vienna • Oxford

Dennis D. Waskul

SELF-GAMES
AND BODY-PLAY

PERSONHOOD IN ONLINE CHAT AND CYBERSEX

PETER LANG
New York • Washington, D.C./Baltimore • Bern
Frankfurt am Main • Berlin • Brussels • Vienna • Oxford

Library of Congress Cataloging-in-Publication Data

Waskul, Dennis D.
Self-games and body-play: personhood in online chat
and cybersex / Dennis D. Waskul.
p. cm. – (Digital formations; 9)
Includes bibliographical references and index.
1. Computer sex. 2. Man-woman relationships. 3. Interpersonal relations.
4. Online chat groups. 5. Internet–Social aspects. I. Title. II. Series.
HQ23 .W27   306.7'0285–dc21   2002070487
ISBN 0-8204-6174-1
ISSN 1526-3169

Die Deutsche Bibliothek-CIP-Einheitsaufnahme

Waskul, Dennis D.:
Self-games and body-play: personhood in online chat
and cybersex / Dennis D. Waskul.
–New York; Washington, D.C. / Baltimore; Bern;
Frankfurt am Main; Berlin; Brussels; Vienna; Oxford: Lang.
(Digital formations; Vol. 9)
ISBN 0-8204-6174-1

Cover design by Dutton & Sherman Design

The paper in this book meets the guidelines for permanence and durability
of the Committee on Production Guidelines for Book Longevity
of the Council of Library Resources.

© 2003 Peter Lang Publishing, Inc., New York
275 Seventh Avenue, 28th Floor, New York, NY 10001
www.peterlangusa.com

Printed in the United States of America

In memory of Mark Allen Douglass

❖ ACKNOWLEDGMENTS

Between 1994 and 2002 I shared my research and scholarship with Mark Douglass. We developed our key ideas together, often over gallons of coffee and beer, depending on the time of day. We spent countless hours writing and critiquing one another's work as we crafted fragmented thoughts into ideas, ideas into papers, and papers into published manuscripts. Together we traveled and presented our work at professional conferences, often exploiting those travels to dine in fine restaurants, socialize with like-minded colleagues, enjoy each other's company, and deliberate the merits of our next research projects. In both our professional work and personal lives we were bound together by a rare bond of profound friendship. Mark was graciously reviewing an early draft of this book in the spring of 2002 when on May 14 a tragic automobile accident claimed his life.

As a coauthor Mark had a huge influence on my work. Mark was my "first editor"; whether working alone or in collaboration I never published a word that didn't first receive his seal of approval. As a professor of sociology Mark was inspiring; I have never met a more gifted teacher, nor one more committed to student growth and development. As a friend Mark is irreplaceable; with all due respect to his real siblings, he was a brother to me in all the ways that matter the most. Mark Douglass was a person of extraordinary wit, intelligence, talent, and compassion. For a man who died at the young age of thirty-two, he left behind an indelible mark; his legacy will be carried on by those whose lives he touched. Mark Douglass is sorely missed and this book is dedicated to his enduring memory.

I owe an enormous debt of gratitude to Charles Edgley. It is doubtful I could have gone very far without Chuck's support and encouragement in times of personal and professional need. His creative influence has permanently altered the nature of my research and his insightful reviews continue to shape my work for the better. I must acknowledge the substantial influence of Kathy Charmaz; I have benefited greatly from her countless hours of tireless effort to improve the quality of my scholarship. Special thanks also go to Steve Jones. Finally, I am especially grateful for the support of my wife

Michele. Not once did she complain as she made enormous sacrifices while I spent years moving from one unstable, under funded, and/or temporary academic position to the next. Not once did she misinterpret my commitment to teaching and scholarship. Not once did she question my motives for spending hours upon hours studying kinky online sex. Of course, all of these things are quite trivial. In the end I am far more grateful for other things she has brought to my life, but these are much too personal to mention here.

❖ TABLE OF CONTENTS

❖ CHAPTER ONE
Everyday Life through the Lens of the Internet

Will the Internet have a transformative effect on society, culture, social institutions, and human relationships? For most people such a question is rhetorical (at best) and pointless (at worst)—either way, easily dismissed as both silly and palpable. The obviousness of the answer locates the query in a list of other equally flaccid yes-no questions, somewhere between "Did Bill Gates become a rich man?" and "Did Bill Clinton have sexual relations with 'that woman'?" We all know the answers to these questions; indeed, they are not questions at all.

Still, it is worth remembering that just one decade ago a question like this was occasion for heated debate. Now, there is little reason to either ask the question or debate its possible answers: the transformative potential of the Internet has already been realized; in a few short years we have witnessed the changes it has made in education, business, commerce, politics, correspondence, and an endless array of other social institutions and experiences of everyday life. Through the 1990s, as the social, cultural, and institutional world got progressively "wired," we were all participants and spectators in the sometimes better, sometimes worse; sometimes obvious, sometimes subtle; sometimes surprising, sometimes predicable; and always varied changes that resulted from this restless wiring.

No one seems to debate whether the Internet is related to various kinds of social and cultural change, and I surely will not revisit those already nostalgic deliberations. Instead, I wish to address other questions—ones that continue to be asked—and along the way entertain possible answers that are not nearly so simple. To frame it plainly by way of contrast, questions concerning whether the Internet is (or has been) transformative may be dismissed as obvious; questions about the meaning and significance of that change cannot be. In other words, while few people deny that the Internet has changed various aspects of society, culture, social institutions, and communication, far fewer seem able to detail what those changes mean, why they are significant, and how these changes might inform us about the experience of life in contemporary society. It is questions of this order that this book seeks to explore.

As anyone can see, the Internet has introduced a host of new communicative possibilities to the experience of everyday life. E-mail, bulletin boards, listserves, webcams, and online chat all represent new ways for people to communicate with others. A large proportion of us use at least some of these channels of communication at least some of the time. Since many have already been integrated into the normative experience of life in society, there are ample reasons to believe that they are not a new and trendy communicative fashion; they will remain a part of our collective communicative universe. Furthermore, by use of these Internet based communication technologies we meet or keep in contact with others. Some of these "Internet people" we know personally while others are virtual strangers—people we only know on the Internet or would not know were it not for the Internet. These new forms of communication bring us into contact with others in new ways, sometimes resulting in surprising outcomes. Consider, for example, the following:

A woman from Fort McMurray, Alberta, left her family after falling in love with an American she met in online chat. Later, she returned to her husband after he learned to correspond with her by e-mail (Dobrovnik 1998).

As the example of the woman from Alberta illustrates, people do meet on the Internet where it is possible to fall in and out of love. Thus using the Internet for the explicit purpose of matching people of like interests seems all too reasonable. Indeed, online dating services have achieved a great degree of popularity. Some people even bypass commercial dating services altogether, creating websites that function as personal advertisements. A twenty-seven year-old man from Virginia created one such site in May of 2000, though his website is truly unique. Using his uncanny resemblance to traditional depictions of Jesus Christ, this man created an elaborate website where he markets his unique appearances in an effort to attract ladies. At his website, entitled "Jesus Seeks Loving Woman," females are invited to an "offering of spiritual togetherness" and a sharing of "tender cuddling that transcends time." Women interested in his "spiritual offerings" may download "sermons," look at pictures of "Jesus," and apply for a date. But interested ladies ought not fret over their spiritual insecurities: "in most cases Jesus will be available and eager to speak to you about spirituality if desired. If you are not spiritual, Jesus will share a beer and pleasant existential banter." Best of all, "young women interested in bathing with Jesus can now have their dream come true. Not only will you make a new friend, but you will be supporting good hygiene and benefiting the environment by conserving water. There are no strings attached except that a picture of us, suitable for family viewing, will be taken and placed on this website as a lasting tribute to our determined efforts at cleanliness" (www.jesus.com)

Of course, "Jesus" isn't the only one who has discovered the enormous potential of posting personal pictures and video on the Internet. The unique appearance of "Jesus" has brought him considerable attention (his website appeared in *Maxim* in March 2001, *Wired* in April 2001, and a variety of other publications), but he is not alone in his extraordinary use of his own self-image on the Internet. In fact, using inexpensive webcams, a woman gave birth in front of a potential audience of millions on the Internet; various "college students" prance about their "dorm rooms" broadcasting their mundane daily routines to the cyberworld; a pair of eighteen year-old "virgins" had sex for the first time before the prying eyes of an unknown number of cyberpeeps; and 1960s drug advocate Timothy Leary contemplated committing suicide live on the Internet (Abate 1998; Krivel 1998).

It is a mistake to assume that all or even most people posting images of themselves on the Internet are seeking an audience of cyberpeeps for extreme (or extremely mundane) events. In fact, at hotornot.com, regular everyday folks post pictures of themselves. Viewers access the site free of charge and rate the attractiveness of these men and women on a scale of one to ten. Ten if they are "hot," one if they are "not," and two to nine for those whose appearances fall between. Once a viewer has rated the attractiveness of one person, he or she is immediately shown the mean score (in case you were wondering if your standards of "hotness" and "notness" corresponded with those of others), and given another image to rate. Viewers continue to rate the attractiveness of one person after the next. Those that post a picture of themselves can quickly find out if, in fact, they are "hot" or "not," and often there are thousands of votes to support that verdict—for better or worse!

Although the Internet can be used for all kinds of communication, it remains a distinctive disembodied context for social interaction. There are lots of things on the Internet, but co-present corporeal bodies have yet to be included. Still, this lack of physical body contact has not hindered the curious, adventurous, and imaginative from having sex "on" the Internet. Indeed, opportunities for sex on the Internet are abundant, so much so that in *Virtual Spaces* Cleo Odzer (1997:43) describes her online sex life by saying that "with the freedom to be and do anything, I had sex with three men at once. I had sex with a woman. I had sex with three men and a woman once. Posing as a man, I had sex with a woman. Posing as a gay man, I had sex with a man. Posing as a man, I had sex with a man who was posing as a woman. I learned all about S & M, as the sadist and as the masochist. I had all sorts of sex in every new way I could think of."

Not all these people having all this sex are necessarily single; many are in committed relationships or marriages. This being the case, numerous psychologists, marriage counselors, and advice columnists have raised predictable concerns about "virtual adultery." In one highly publicized case, a New Jersey man filed for divorce, accusing his wife of carrying on a "virtual affair" via computer with a cybersex partner whom she met on America Online and knew only by his screen name, "the Weasel." The divorce papers included dozens of e-mail exchanges between his wife and "the Weasel," who turned out to be a man who was also married (Associated Press, Feb. 2, 1996).

If having sex on the Internet isn't enough, one may purchase highly sexualized items or objects that have been *used* for sexual activities. Numerous websites sell lurid commodities such as used panties, used vibrators, and a wide assortment of other similarly soiled materials. At one website, shoppers can find bargains such as a pair of panties worn for two days and nights for ten dollars, fifteen dollars for panties worn four days and nights, twenty dollars for a bra and panty set, five dollars for dirty socks, one dollar for a lipstick imprint on paper, and five dollars "for pussy lip wet cum print on paper."

Those who believe that sex and pornography are the most disturbing things on the Internet have obviously never visited Rotten.com. Although the website has many features, its primary attraction is a huge assortment of some of the most gruesome and explicit images of human death and disfigurement imaginable—often brought on by traumatic and violent means. Rotten.com brings the theater of death to a whole new level by featuring graphic images of people who have died in motor vehicle accidents, children who have been brutally murdered, dismembered body parts, the results of a successful suicide by shotgun blast to the head, an autopsy of an infant, a body torn to pieces by a pipe bomb, and much more for those who can stomach these truly disturbing images.

As many people have discovered, the Internet can prove useful for broadcasting to "the world" what kind of people we are and the experiences we have encountered in our travels through life. After all, that would seem to be the primary use of personal websites. Some people have supplemented these personal websites with online diaries and similar kinds of biographic editorials. In his explicit online diary entitled "to daze," Swarthmore college student Justin Hall reveals, among many things, his experiments with drugs, sex, and how he contracted and treated a sexually transmitted disease. A woman known online as FauveGrrl maintains a popular web page appropriately called "The List." It is a list of the thirty-eight men she has slept with. "The List" identifies names (some fictional), the date they had sex, and "notes." Notes include such passages as: "He was the oldest stepbrother of one of my girlfriends. He was too drunk to perform", and, "He was a stuntman. I slept with him on the night I was supposed to have been married to Theodore. He had the best body I'd ever seen."

In 1982, "Julie" signed on to CompuServe, describing herself as a New York neuropsychologist who was crippled from an auto accident that killed her boyfriend. She also described herself as a bisexual atheist who occasionally smoked marijuana. "Julie" started an online women's group, and many participants found her advice helpful. "Julie" became popular and well liked among her cyberpals. However, when it was discovered that "Julie" (she) was really Sanford (he), "her" previous friends felt emotionally violated. One person said, "I felt raped. I felt as if my deepest secrets were violated. The good things Julie did...were all done by deception" (Stone 1995; Van Gelder 1985).

Finally, the Internet has produced new issues for the law involving distribution of pornography. The operators of adult bulletin board "Amateur Action" were given a jail sentence in *Tennessee* for distributing obscene pictures from their home base in *California* (Rose 1995).

What are we to make of all this? It's hard to say for sure. Situations like these range from the amusingly playful to the disturbingly bizarre. But one thing is for certain: each of these situations refers to unforeseen social, cultural, and legal implications of the Internet; each is indicative of certain shifting boundaries of personhood and social interaction evident not only on the Internet, but also observable elsewhere in society.

"There is little doubt that the Internet, for all its faults, is perhaps the most fascinating and explosive technological and social development of the twentieth century" (Whittle 1996:15). Nor is there much doubt that the Internet is related to a significant degree of change—and that should not be surprising. The Internet is a new technology and all "new technologies alter the structure of our interests: the things we think *about*. They alter the character of our symbols: the things we think *with*. And they alter the nature of community: the arena in which thoughts develop" (Postman 1992: 20). It is therefore easy to conclude that "the Internet has grown in recent years from a

fringe cultural phenomenon to a significant site of cultural transformation
and production in its own right" (Porter 1997: xvii).

In contrast to the controversy of a decade ago, it has become common—
even fashionable—to claim that the Internet has had a transformative effect
on society, culture, institutions, communities, social interaction, and the experiences of personhood. However, what kind of changes are we talking
about? How can we describe the effects of these changes? On what? Are we
to believe that the Internet is the source of all this change, or has society
changed due to other causes—ones that are independent of the Internet?
Does the Internet bring about changes in the human landscape, or does it
merely reflect the kind of people and society we had already become or have
always been? Is the Internet a site for entirely new social experiences, or
does it simply reconfigure old ones? I think, on one hand, we are wise
enough to recognize that whatever changes have occurred in relationship to
the Internet, they are not mono-causal, entirely positive, or entirely negative,
nor can they be described as either one thing or another; we are sensible
enough to acknowledge that this kind of social change is too complex for
simple binary questions and equally simple answers. But, on the other hand,
to ask the question again in a slightly different context: do we have any sense
of what to make of all this? Here, like before, it would seem that careful consideration invariably produces an uncannily similar answer: It's hard to say
for sure. And, I'm not afraid to add, I cannot say for sure myself.

I suppose, since I am a sociologist who has spent years studying certain
dimensions of the Internet, many would expect that I have a better answer.
Those familiar with sociology might be inclined to point out that social
change is central to the subjects and theoretical foundations of the field—a
point I will neither argue nor deny. Surely, there are theoretical models in
sociology that might prove useful in retrospectively understanding these
changes and perhaps even providing clues about future ones. Even so, in
spite of considerable study and empirical investigation, I have to admit that I
am not privy to any special, secret, or otherwise undisclosed knowledge that
I intend to detail in this book. Furthermore, if such knowledge exists, I remain utterly ignorant if for no other reason than my attention has been directed elsewhere. Frankly, I have taken from sociology something else; by
which I mean to say I have peered into the Internet for other reasons, asking
different questions about its significance in our lives.

Musings about the Internet in relation to the nature, extent, and direction
of social change have been the interest of others, and will remain so. My
primary interests have always been locked on something else, and that too

will remain so. I am convinced that one of the most important and fascinating dimensions of the Internet is not so much how, why, or to what extent it has been a catalyst for change, but instead what these experiences can tell us about the kind of people we are. Approaching the Internet in this way has compelled me to think of it differently. The Internet may, in fact, be a source of significant social change, but I am much more interested in the Internet as a *site* for human relations. The Internet is certainly a medium of communication, but I am also interested in how it constitutes a unique *environment* for social interaction. The Internet may function as a technological plaything for those who enjoy the privileges of access, but I wish to understand it as a *context* for human experiences of self and others. In short, I principally conceive of the Internet as a site, environment, and context in, on, and through which we can experience others and ourselves in ways that give new meaning to old questions.

Even more, I perceive these sites and experiences pragmatically in the sense that I am interested in their relationship to the experience of everyday life and our understandings of that experience. This kind of pragmatic approach provides us insight into the nature of personhood on and *off*-line; the latter of which interests me the most. In short, this book has everything and nothing to do with personhood on the Internet. This book is all about examining the Internet, but only to the extent that we can see *through* it; positing the Internet as a lens by which we may better see, understand, and contemplate the nature of personhood in everyday life.

Social Interaction and the Internet:
Play and the Sensual Pleasures of Conversation

While early scholars of computer-mediated communication were preoccupied with the effects of the Internet on work, business, and other institutional environments, I was impressed with how people were using it as a form of leisure for the purposes of entertainment and communication play. Studies focusing on the impact of the Internet on institutional environments are important. However, my work has always sought to explore how the Internet is used and experienced in more casual contexts that differ significantly from its functions in business and bureaucracy. Millions of people use the Internet for leisure, entertainment, and play—a large portion of which involves interaction with others in unique ways—and, I would venture to say, it is these uses that principally attract people to the Internet in the first place.

David Porter (1997:xii) could not have been more correct when he wrote that "what continues most powerfully to draw people to the Internet is its power and novelty as a medium of person-to-person communication." While a major portion of this person-to-person communication is for professional purposes, another large fraction thrives on the seemingly forgotten and easily ignored pleasures of conversation—that literally wonder-full form of casual social interaction where the sometimes irresistible draw is linked to its intrinsic promise of satisfying communicative leisure. That is, after all, what online chat is all about: conversation for the delight of conversing, communication for the joy of communicating, interaction for the sake of interacting, written talk for the enjoyment of writing and talking with others. Thus, in online chat, having something to say is often far less important than simply saying something. And, like any other kind of conversation, the pleasures of having said something are realized when others take interest in what we have said, and we return their attention in an ongoing, creative, artful, and expressive communicative dance. To spend time in online chat is to learn these pleasures firsthand; to understand these forms of interaction is to recognize the importance of these delights in our life firsthand.

Even more, online chat often bustles with unabashed flirtation, sometimes involving candid and explicit sexual interactions. While surprising to some, there are understandable reasons for this kind of shameless flirtation and sexual interaction. First, sex provides a ready-made topic for conversation simply because everyone is doing it, thinking about doing it, or wishing they were doing it. Most everyone has something to say about sex and romance, their experiences with it, fantasies, desires, and so on. In online chat, people can *always* flirt and talk of sex—*especially* if they do not know each other, share little else in common, don't have anything else to talk about, or even find one another otherwise uninteresting. Furthermore, the fact that coquetry abounds in online communication environments should not be surprising if for no other reason than it is equally abundant in everyday face-to-face interaction. Flirtation is common on the Internet, just as it is common in bars, schools, work environments, parties, dance clubs, and just about anywhere people find themselves in the presence of others. Granted, the coquetry of everyday face-to-face interaction is typically much less blunt or overt than that which is found in Internet environments, but even so, one can hardly deny its significant presence in both contexts. Nor can we deny knowing (first- or secondhand) people in "real life" with whom we have formed entire relationships on the basis of sex and coquetry, in spite of sharing little else in common—not unlike many online relationships. Certainly, sex is a convenient, ready-made, and common topic for communication on the Internet, but so too is it in everyday life.

Second, we ought not overlook the interesting similarities between the pleasures of conversation and sex. These similarities make them an especially savory combination in the experience of flirting that is particularly well suited to online communications. In many respects the pleasures of conversation are quite similar to the pleasures of sex. In both sex and conversation we give and receive; are attentive and tended to; we reveal as something is revealed to us; we are explorer and explored; the outcome is never completely predictable, but the experience is still made orderly by unwritten and generally shared rules; turns are taken; there are skills and techniques to be learned and developed; one risks being hurt and hurting others; there are rewards of being pleased and pleasing others; we may misunderstand or be misunderstood; being good at it means that we can con and be conned by others; and conversation, like sex, can leave us gratified, exhausted, excited, disappointed, worn-out, or longing for more. Sex and conversation share much in common. However, I'd rather not stretch this metaphor too far; to a certain extent sex can be likened to almost anything (conversation, war, an economic transaction, work, etc.) and almost anything can be likened sex (conversation, war, an economic transaction, work, etc.). This metaphor means everything and nothing at once. Even so, it is still worth considering the extent to which sex represents a certain paradigmatic form of social interaction with parallels to conversation that are most directly realized in the innocent and sometimes not-so-innocent experience of flirtation. Thus, if we think about it enough, it is not surprising that strangers who meet for conversation in the liminal ether of electronic space wind up flirting, sometimes quite overtly—in fact, it would be far more surprising if they did not.

These forms of interaction on the Internet are uniquely social and they are the subjects of this book.[1] These are social experiences not only because they are distinctively creative, expressive, and playful, but also because they have a special emergent quality that represents the "stuff" of sociological analysis. That is, online interactions may be patterned and influenced by a wide array of variables, but those patterns and variables do not predetermine outcomes. Instead, participants themselves, in an ongoing process of communication, extensively shape the nature of these interactions and their experiences within it. On the Internet, participants interact through channels of communication by which they collectively construct a social context and themselves within it. Yet, those online social constructions also reflect back upon them in new and novel ways that can have their own indeterminate, yet still perceptible shaping influence. This process, as it occurs online, is not fundamentally different from everyday life. In fact, this normative process represents the fundamental core of society itself. It is the various dimensions of this relationship, as it occurs in the playful experiences of online chat and cybersex, that this book seeks to explore, and I will add to this a special interest in the implications of these experiences for better understanding the more general nature of personhood in society.

Antienvironments:
Peeping through Doors of Perception

At a minimum, the circumstances described earlier illustrate the extent to which online social environments constitute a realm of experience that is outside of the ordinary. They are extraordinary, in the literal sense of being extra-ordinary. Whether it is "Jesus seeking a loving woman," or a "live" peep into the "dorm room" of a "college student," or a spouse caught red-handed (figuratively or literally) in a torrid love affair with someone they have never met—these circumstances are, at the very least, peculiar. And it isn't too much of a stretch to suggest that only on the Internet do these kinds of peculiarities make sense, if only in some strange way. After all, if Cleo Odzer (1997: 43) were to tell us that "posing as a man, I had sex with a man who was posing as a woman," we would think she was insane, very involved in the transsexual/cross-dressing scene, or had a hearty taste for unusually imaginative sex. Regardless, her description would be difficult for most of us to comprehend. However, once we realized she was describing her *online* sex life, her description would become rather unsurprising; we could nod our head in understanding—it would make sense, if only in some strange way. Indeed, the Internet makes possible, and even understandable, experiences that in any other context would be unusual or just plain bizarre.

On the Internet, we find spaces and social places where forms of personhood are transformed into symbolic constructs that are not necessarily grounded in or referenced from either the empirical world or normative expectations of society upon individuals. These social worlds and forms of selfhood are fundamentally different from everyday life for numerous reasons, not the least of which are the obvious and significant alterations in the means by which they are produced and experienced. It is for these reasons that these kinds of experiences can be approached and understood as a kind of "antienvironment" (McLuhan and Parker 1968) where commonplace occurrences are experientially transformed into something different, unusual, atypical, or otherwise unlike their original or normative form.

Erving Goffman (1974:564) once noted that these kinds of extra-ordinary experiences are especially revealing and of unique significance for those who wish to understand the nature of everyday life in society. In these antienvironments, one can observe with utmost clarity characteristics of everyday social life that are otherwise quite difficult to perceive:

> Realms of being other than the ordinary provide natural experiments in which a property of ordinary activity is displayed or contrasted in a clarified and clarifying way. The design in accordance with which everyday experience is put together can be seen as a special variation on general themes, as ways of doing things that can be done in other ways. Seeing these differences (and similarities) means seeing. What is implicit and concealed can thus be unpacked, unraveled, revealed.

As Goffman suggests, these extra-ordinary experiences have a special illuminating potential for understanding personhood, our social world, and relations between the two. These antienvironments provide a unique opportunity to see clearly through open doors of perception and notice things that might otherwise be overlooked. Examining these experiences potentially heightens our awareness of previously unnoticed differences and similarities. These kinds of experiences therefore present ideal conditions for both analysis and understanding.

Online social worlds represent one such antienvironment where everyday occurrences assume new meaning. As we shall learn in the chapters to come, the "dynamics that occur in RL [real life]...manifest themselves more explicitly in cyberspace. So subtle are these dynamics in real life that people aren't even aware of them" (Odzer 1997: 4). Extraordinary circumstances allow us to be made aware of those dynamics and, as Goffman suggests, provide a valuable lens through which we may better assess their meaning in new ways that hold a special promise of illuminating processes central to the experience of life in society. These technologically sustained "antienvironments" open the door of perception to people otherwise numbed in a non-perceivable situation (McLuhan and Parker 1968). This book intends to peep in through those doors of perception; in this way, the subjects discussed in this book do not merely involve social games people play with computers. Instead, these games provide temporary and isolated glimpses into the shifting boundaries of the relationships between self and society made manifest in computer networking technologies.

Self, Social Interaction, and Cultural Storytelling:
Some Words on Theory and Method

This book posits the Internet as a lens through which we may better understand the nature of personhood in contemporary society. Therefore, our primary task is a conceptual one, and a dual one at that: to describe, explain, and understand experiences of personhood in online chat and cybersex, and to entertain questions about the significance of those understandings for the ways we describe, explain, and understand personhood in everyday life. Such an approach is, of course, an exercise in theory.[2]

This analysis and discussion stands on the shoulders of well-established theoretical models in sociological social psychology and also draws significantly from a more recent body of loosely associated literatures that are most commonly called "postmodern." Presently, it is not necessary to detail these theories. Instead, these theoretical understandings shall be explained as they are applied to specific circumstances in our unfolding analysis of online chat and cybersex throughout this book. Nevertheless, a brief introduction is in order.

Analysis and discussion proceeds primarily from the interpretive frameworks of symbolic interaction and dramaturgy. Symbolic interaction is a theoretical approach that focuses squarely on the important role of symbols and meaning to the experience of life in society. While any animal can be said to act or interact (in the loose sense of the words), only human action and interaction (in the proper sense of the words) is mediated by symbols and meaning. Humans act toward things on the basis of what things mean, and thus, what things are is far less important than what things mean. Since the world does not come prepackaged in intrinsically meaningful categories, these symbolic constructs are inherently social; they are emergent within social and cultural systems, understood by reference to broader interpretive frameworks, and learned in an extensive process of socialization.

Symbolic interaction focuses on and to a certain extent is preoccupied with two major interests that are of special utility for understanding the nature of online chat and cybersex. First, symbolic interaction emphasizes the important role of language to society and social interaction. Language is a complex system of significant symbols that perpetually mediates and structures how we define, interpret, and act upon the world. "All other symbolic systems can be interpreted only by means of language.... It is the instrument by means of which every designation, every interpretation, every conceptualization, and almost every communication of experience is ultimately accomplished. What is not expressed in language is not experienced and has no meaning" (Hertzler 1965:29). While symbolic interactionists must frequently argue a case for the central role of language in the experience of life in society, such arguments are largely unnecessary for the majority of online interaction. As anybody can see, social interaction on the Internet is almost entirely accomplished through language—overwhelmingly, written language. To interact in an online chatroom or engage in text cybersex, it is quite necessary to communicate the whole of the experience through typed words. Thus, symbolic interactionists may be correct when they claim that language always mediates our experiences of life in society, but it is almost impossible to deny the central, obvious, and paramount significance of language to the experience of social interaction on the Internet. Symbolic interaction provides a coherent framework for fully understanding this relationship.

Second, symbolic interactionists are preoccupied with human experiences of self. Symbolic interaction provides a sophisticated framework for understanding how individuals accomplish, maintain, and transform self in society. Yet, oddly enough, this is the one concept that symbolic interactionists struggle the most to explain in simple, neat, and terse terms. One can be endlessly amused by simply asking symbolic interactionists to briefly explain "self"; one can read volumes of symbolic interactionist analyses of self and never find a definition. However, the tortured struggle of symbolic interactionist attempts to explain selfhood in straightforward and uncomplicated terms is not due to insufficient thought or poorly conceived ideas. To a symbolic interactionist, such a task is impossible—selfhood is not straightforward, or uncomplicated, and it cannot be explained in simple, neat, and terse terms without seriously undermining its essential character.

In symbolic interaction, "self" is not synonymous with ego, personality, identity, or even the person. Like anything else, self is symbolic; it is a meaning, not a thing. For this reason, self is not something that we "have." To a symbolic interactionist few things are more curious than the questionable assertion that we "have" a self or "personality," as if these "things" were a possession carried about in our pockets or an anatomical appendage to our body. Self is not something that we "have"; instead, it is something that we communicate.

Yet even this is far too simple, for selfhood is not something that emulates from within the individual radiating out to the world from the singular source of our body (or any component of it, such as our mouth or brain). From a symbolic interactionist perspective, one cannot have a self all by one's self. Instead, selfhood is something that emerges *between* people in situated contexts of social interaction. For this reason, self is something that is not only communicated; it is something that is interactively crafted and necessarily dynamic. Self is elastic. A self is fashioned moment by moment, situation by situation, as we are one thing to one person and something else to another. It is in this respect that a multiple personality is normal (Mead

1934); the answer to the question "Who am I?" cannot be answered in the singular; it is necessarily plural; it is necessarily a list.

Symbolic interactionist accounts of self are especially well suited to understanding the dynamics of personhood on the Internet. On the Internet, self is clearly something that is not contained within individuals but communicated among people, largely through typed words. Furthermore, as most people know (and will be explored more fully later), self on the Internet can assume a stunning range of expression. On the Internet, a man may be a woman; a bored housewife may be a teenage boy; "Julie," the woman who established the online women's support group, may actually be Sanford, the guy playing "Julie." Online social interaction fully illustrates the symbolic interactionist assertion that self is a symbolic system of meanings that are communicated from one moment to the next. In this process, the symbolic interactionist assertion that multiple personality is normal is also fully realized in online social environments. Therefore, it is from the framework of symbolic interaction that we shall gain insight into these dynamics of online chat and cybersex yet also clearly see and explore their parallels to routine off-line experiences of life in society.

In addition to symbolic interaction, dramaturgy also proves central to this analysis. Indeed, constant reference will be made to work of Erving Goffman, who is perhaps most prominently associated with the dramaturgical tradition in sociology. Symbolic interaction and dramaturgy are, for all practical purposes, the conceptual equivalent of identical twins—they are more alike than different, yet unique in their own subtle ways for those who know them best. Dramaturgists would not disagree with any of the points I've briefly explained about symbolic interaction (although they would insist that I expand my definition of "language" to include both discursive and non-discursive forms of communication, but that is a small matter). The key difference is that dramaturgists focus more intently on how social reality is something that we *do*.

In the words of Shakespeare, "All the world's is a stage and all the men and women merely players." Dramaturgy borrows this theatrical metaphor to better understand how we "do" a society and thus accomplish meaning in our life. Which is to say that in dramaturgy meaning does not precede social interaction; instead, it is established with it. This analytical focus is the critical lesson to be taken from dramaturgy in our attempts to understand online chat and cybersex. On the Internet, characteristics often considered an innate part of being a person are necessarily revealed to be performative dramas. People are who and what they *claim* to be, and the reality of those claims boil down to how well the individual *plays* at being that kind of person. Dramaturgists have always insisted "that without a presentation of self, a self is not possible" (Krielkamp 1976: 137), and interaction on the Internet makes this readily apparent. In online communication, things like race, age, gender, and other important symbols of selfhood are not nouns—they are not qualities of a person. Instead, they are verbs—they are something that we *do* in a process of communication. Qualities of personhood are not static constants that define who we are; they are, instead, symbolic identity utensils that we use to fashion a self.

In addition to symbolic interaction and dramaturgy, my analysis is also informed by postmodern literatures. As with everything in the fuzzy world of the "posts," it is difficult to pin down the nature of the postmodern condition, concepts, and theoretical accounts concretely. However, at the very least postmodernists claim that the era of "modernity," which included positivism in the social sciences, a belief in human "progress," and a generally scientific and exacting view of the world, has ended. The postmodern condition is characterized by a deep sense of relativity, discontinuity, and interpretation replacing "facts." Thus, uncertainty. It involves the demise of ultimate truths, and therefore, all forms of authority are called into question. It is a condition in which reality is increasingly indistinguishable from representations of reality.

Postmodern selfhood reflects the very same qualities of the postmodern world: it is depthless, transient, radically multiple, hyper-fluid, and decentered. The postmodern self is ultimately rooted within the shifting tides of unstable, ever-changing social relations, often made possible by an expansive network of communication technologies. This postmodern emphasis on the depthless and transitory nature of self and society has a special resonance for online social interaction. On the Internet the whole of one's self and experiences of social interaction resides in fleeting words communicated in an electronic space without place. Indeed, there is a certain postmodern nihilistic relativity that pervades online social environments, where the distinctions between truth and lies are either indistinguishable or irrelevant, distinctions between reality and representations of reality are immaterial, and anything goes in a no-holds-barred orgy of communication. In our analysis of online chat and cybersex, these parallels between social interaction on the Internet and the postmodern condition shall not be ignored.

While our task is primarily conceptual, it is not entirely. This analysis is empirically grounded; it proceeds through the methods of ethnographic research. That is, data was collected, analyzed, and discussed from the perspective of participants in online chat and cybersex. Furthermore, the contents of this book are the product of three related ethnographic studies, each seeking to describe, explain, and understand different dimensions of the experience of personhood in various venues of online social interaction. The first study focuses on online chat and the experiences of self in these unique Internet environments. The second examines text-based cybersex, exploring relationships between self, body, and social interaction in circumstances where the body cannot be seen (or otherwise interacted with) by others. The third study turns its attention to televideo cybersex, where participants interact through digital cameras, allowing bodies to be seen while selves are communicated. Each of these studies examines different dimensions of personhood in online chat and cybersex, and this book unites the trilogy.

A wide variety of research methods were used in the processes of data collection. I will not belabor the reader with tedious discussions of these methodologies (those for whom methods are of interest or importance can refer to the appendix). It is sufficient to specify that while several methods of data collection were used, the majority of data came from a total of 152 open-ended real-time e-interviews with online chat and cybersex participants.

By any standard of ethnographic research, 152 interviews would qualify as an ample volume of data for significant, valid, and reliable analysis. However, I am not preoccupied with the rather trivial matters of "scientific" "validity" and "reliability." Quite simply, my purpose in these studies is to weave narratives about human behavior; to tell empirical stories that I believe are important to conceiving the nature of personhood on the Internet and better grasping its relevance to understanding significant dimensions of human social life in general. Granted, in contrast to pure fiction, the ethnographic stories that I tell are based within the findings of my research. Nonetheless, it is fair to conceive of my work as a variant form of storytelling—a conception that may apply equally as well to any other product of the social "sciences" (see Postman 1988).

As a cultural storyteller, I believe that the general purpose of social research is not to discover something new; instead, it is to "rediscover the truths about social life; to comment on and criticize the moral behavior of people; and finally, to put forward metaphors, images, and ideas that can help people live with some measure of understanding and dignity" (Postman 1988: 18). In this way, good social research is a form of storytelling that has a moral purpose. Good social research tells us stories that we ought to hear and puts forth understandings that are worth knowing or considering in the life that we live. Like religious parables, good social research purports to tell us something important about our life, the lives of others, and the relationships between them.

My research has been guided by the intent to craft these kinds of narratives about human behavior on the Internet. Furthermore, the stories I tell tend to adhere to one central and fascinating moral theme—each is about people struggling to understand themselves and their world using whatever materials they have at hand. I believe that these are dramatic and gripping stories. To aid in constructing these narratives, I use ethnographic data as a mechanism for discussing issues that I believe are important for understanding the struggles of personhood in contemporary society. Ultimately, however, I do not believe my studies qualify as "science"—they are simply narratives. Thus, in final analysis, I believe that the value and utility of my research boils down to one simple but important question: Will readers find my narratives believable, enlightening, or thought provoking? I cannot know for sure, but it is my hope and intent to have crafted empirical narratives that accomplish precisely these things.

NOTES

1. Some may wonder why I have avoided data collection on MUDs (and various other related forms of Multi-User Domains). I have avoided these contexts for research (not necessarily for play) for two main reasons. First, there is a substantial amount of information regarding MUDs already available and comparatively little that focuses directly on online chat and cybersex. Second, MUDs are, at least to some extent, game-oriented social environments. Game-oriented real-time communications are at least influenced, if not structured, by the explicit and implicit rules of the game. They are by nature fantasy or role-playing activities, and being in character is part of the game. Furthermore, the social worlds that emerge in these contexts are at least partially determined by the context of the game itself. Although participants in these games can construct their own contexts for play, for all practical purposes the social world is predefined—a prepared context in which persons can role-play for the sake of the game. For these reasons, game-oriented real-time communications are of less interest to me than unstructured chat and cybersex forms of real-time communication.

2. I will be the first to admit that there are various subjects explored in this book that are impossible to discuss without tainting them with personal bias. For example, one cannot write honestly or meaningfully about kinky online sex without making one's biases explicit. That being the case, I will also admit to knowing fully what my biases are. Thus I can state with certainty what this book is *not*. Topping that list, this book is not a manifesto of moralizing claims about the "dangers" and "addictions" of the Internet and cybersex. There are enough grim-faced moralists already prowling the halls of the social sciences and an ample selection of their books and articles already available for those who share their taste for this kind of variant branch of theology. I have no interest in join-

ing these efforts to incarnate the Internet as a contemporary secular Lucifer—a seemingly omnipresent evil that lurks beneath the surface of what Internet users think are just fun and games; an omnipotent wickedness whose potentially disastrous influence is beyond our power to control without the aid of saintly paladins to rescue all of us from our ignorance and addiction(s). To phrase it politely, I have grown weary of these kind of tiresome arguments. I will not suggest that what I have to say is free of value claims; I have no intent to hide beneath the utterly transparent cloak of "scientific neutrality." Anyone is free to criticize my biases, as I am inclined to criticize others. But even so, I believe that overtly moralistic accounts of online chat and cybersex are embarrassing, misleading, and a serious hindrance to understanding. It is folly to assume that one can meaningfully claim to understand that which one has condemned from the start, and I shall seek to avoid doing at least that much.

❖ CHAPTER TWO

Cyberself: Selfhood in Online Chat *

Who are we when online? What can personhood mean when experienced on a computer? When interacting with others on the Internet, what can be attributed to the technology, the person, and interactions between people? Why do some people believe they are "more themselves" online then in "real life," how do they distinguish between these two (or more) selves, and what are the implications of this? What does it mean when disembodied virtual personae have better conversations, make better friends, or have better sex on the Internet than embodied people in everyday life? Where, exactly, are people meeting, interacting, and doing all these things? Do they occur *on* the Internet? *In* the Internet? *Through* the Internet? There are no simple answers to these kinds of questions, yet each suggests the extent to which computer-mediated communications pose new questions to old problems of social interaction and selfhood.

"Embedded in every tool is an ideological bias, a predisposition to construct the world as one thing rather than another, to value one thing over another, to amplify one sense or skill or attitude more loudly than another" (Postman 1992:13). Karl Marx understood this when he wrote about how technology reveals the ways that humans deal with nature, creating certain "modes" by which people perceive reality that are central to social life in a broad historical context. In other words, Marx was aware of an important relationship between technology and the symbolic universe of human social and cultural activity.[1] Much more directly, Marshall McLuhan (1964:7) fully understood this when he wrote his famous aphorism, "the medium is the message."

McLuhan pronounced loud and clear what Marx only suggested in a hushed whisper: every technology contains a language, grammar, and syntax that influences the content of what that technology is used to produce. This basic truism is perhaps most easily seen in communication technologies

---

* An early version of the contents of this chapter was coauthored by Mark Douglass and published in *The Information Society* (1997, Volume 13, Number 4, pages 375–397).

here, "by changing the boundaries of social situations, electronic media do not simply give us quicker or more thorough access to events and behaviors. They give us, instead, new events and new behaviors" (Meyrowitz 1985: 43). Radio is more than amplified sound. Telephone is more than voice correspondence. Television is more than radio with pictures. The Internet is more than a conveyer of information. Each of these technologies is a *medium;* an environment for communication and interaction that is structured according to its own symbolic code which necessarily alters the functions, significance, and effects of what is communicated. Thus, it is no surprise that contemporary communication technologies are bringing about new questions to old problems of social interaction and selfhood—they are, in fact, bringing about new forms of social interaction and selfhood altogether.

Simply stated, the same content pressed through a medium of communication is not the same at all—a McLuhanesque metamorphosis transforms it into something else. I contend that this process is fundamentally the same for human experiences of personhood as it is for any other content that may be pressed through the conventions of a medium. That is, when we push our self through the conventions of a medium of communication, the same self does not come out the other end; when we push our body through the conventions of a medium, the same body does not come out the other end. In online chat and cybersex, self and body are subject to this kind of metamorphosis, where the medium assures that personhood will be converted into something other than what it is in another mode of communication and interaction.

With this in mind, our task is twofold. Our first task is to understand the nature of this McLuhanesque communicative metamorphosis. Our second task is to explore the implications of these experiences for better understanding the nature of personhood in society. We will begin this analysis by examining online chat, with special interest in experiences of selfhood in these environments. This chapter will detail a conceptual framework for understanding Internet social environments and online interaction, provide an empirically grounded analysis of selfhood in online chat, and conclude by discussing key characteristics that define this form of social interaction.

Cyberspace, Cyberplace, and Social Reality on the Internet

Framing *real* as something different from *virtual* is my conceptual move, not theirs.... I've had to learn not only about how they experience their lives both on and offline, but about how much I was limiting my understanding of them through my presumptions about the term *real*.

—Annette Markham, *Life Online* (1998)

There is no such *thing* as "cyberspace." The word is a concept; an abstraction used to refer to something that isn't a thing at all. To a certain extent, by speaking or writing unproblematically about "cyberspace," we are co-conspirators in definitions that approximate William Gibson's (1984) original articulation of the term: a consensual hallucination that refers to the virtually incomprehensible and seemingly endless array of networked computer mediums, supporting technologies, and innumerable communications that occur by these means.

However, to say there is no such thing as cyberspace is not to say that it is only a hallucination. Instead, cyberspace rightly belongs in a special category of abstractions that have a similar unique basis of reality. This category includes things like society, social institutions, norms, and values—for all practical purposes, the "stuff" of social reality. After all, there is no such thing as a society either. We all live lifetimes in society and none of us have ever seen one. Nor is there any such thing as an institution, a norm, or a value. All we "see" are the doings of people, the consequences of what they have done, and the places where these activities occur. Which is to say, in tandem with our definition of cyberspace, these "things" are concepts, abstractions we evoke to make meaningful the virtually incomprehensible and seemingly endless array of innumerable activities that people do. For this reason, cyberspace—like society—is not something we can see directly, but our inability to do so utterly misses the point: representations of what these things mean are all around us, and these meanings are of utmost significance. Thus, it is the consensual part of our definition of cyberspace that illustrates its important reality and it is this connotation that we ought to emphasize in our understandings of it. As Mark Poster (1997: 205) explains:

> The Internet is more like a social space than a thing; its effects are more like those of Germany than those of hammers. The effect of Germany upon the people within it is to make them Germans (at least for the most part); the effect of hammers is not to make people hammers.... As long as we understand the Internet as a hammer we will fail to discern the way it is like Germany. The problem is that modern perspectives tend to reduce the Internet to a hammer.

From this perspective, the problem with most conceptions of cyberspace is an overemphasis on the "cyber" (the technology), and an underemphasis on the "space" (which is something that is socially produced). Missing entirely from these discussions is any mention of cyber*place* (which is something that is symbolically experienced by participants within these cyberspaces). Thus, we ought to begin by recognizing how cyber and space become cyberspace on the Internet, and how these cyberspaces are made into cyber*places* through communication and human interaction.

Borrowing a concept from a chronically misunderstood sociologist, cyberspace might best be understood as a phenomenon *sui generis;* it is something uniquely endowed with *sui generis* characteristics (Durkheim 1898[1974]). This is to say that cyberspace is something greater than the sum of its parts. Surely, on one hand, the Internet is a series of interconnected wires, cables, electronics, and computer technologies. But even so, on the other hand, these technologies comprise a vast network of mediums and environments for human communication. The social and cultural dimensions of the Internet are emergent from communication and interaction on this latter "other hand." In other words, the Internet is more than the sum of its technological parts; it is a network of unique mediums for communication—"spaces"—where people meet others, socialize, play, correspond, do business, shop, publish creative works, converse, flirt, have sex, and so on. This is also to say that the social environment of the Internet cannot be explained or reduced to the technological components that comprise the medium. One cannot in any meaningful way understand the nature of online chat or cybersex by reference to the cables, wires, or binary codes by which these experiences are made possible. As individuals interact in, on, and through computer networks, a socially constructed environment—a "place"—emerges that is greater than the sum of the wires, telephone lines, digital cables, and computers that comprise it. In short, what emerges is inherently social and circumstantially technological—not viceversa. Thus, technologies (*cyber*space) create mediums of communication (cyber*space*) by which people interact in ways that are made meaningful through social interaction (cyber*place*). It is in assorted configurations of these variable elements of cyber, space, and experiences of place that various dimensions of the Internet acquire their unique character.

In this way, cyberspace is similar to any other social space. It is a socially constructed reality; its unique feature is that it exists within computer networks and supporting technologies. Like other constructed realities, cyberspace isn't so much a physical thing as it is a concept that contains shared meanings, understandings, and information. Approaching cyberspace this way magnifies how it is similar to any other constructed reality. For example, one would be foolish to believe they can understand the nature of, say, church or school by merely examining the buildings, cathedrals, desks, and pews where church and school are experienced. The reality of church and school are emergent from what people think and do—and how those thoughts and doings intersect with the thoughts and doings of others—when they are in those buildings, cathedrals, desks, and pews. The reality of cyberspace and the experience of cyberplace are much more like these other social spaces than different.

This approach to understanding cyberspace effectively repudiates claims that social interaction, self, and community on the Internet are, in some curiously unspecified way, less "real" than any other experience of social interaction, self, and community.[2] This is especially instructive since such claims are common, often found embedded in blatantly moralistic preachings. Consider, for example, this account:

> Some Americans prefer cybersex over real sex, or spend their Saturday nights in Internet "chat rooms" instead of talking with friends face to face. They check out home pages on the Web long after they should have gone to bed. And some people are picking up a computer mouse instead of a remote control.... "We forget the value in our own physicality," says Kerric Harvey, director of the Electronic Media program at George Washington University. "Our bodies and our physical world begin to matter less and less, and the next thing you know we've nickled and dimed ourselves out of relevance. Nothing can, and nothing should ever, replace the real." ..."We unlearn how to be human with each other if we don't practice it." (M. Katz 1998: A2)

I hope just about anyone can see why these claims (and others like them) are unreservedly indefensible. To note that "some Americans" enjoy cybersex hardly justifies the apparent conclusion that these people prefer cybersex to "real" sex (i.e., what we might call "conventional sex"; sex that involves contact with the flesh and fluids of others). Similarly, to say that some people "spend their Saturday nights in Internet 'chat rooms'" hardly precludes significant face-to-face relationships with "real" friends in any (or all) of the other six nights (or days) of the week. To raise concern for people who "pick up a computer mouse instead of a remote control" would be amusingly ironic were the statement meant to be a joke. I am boggled by what Kerric Harvey means when she says "nothing can, and nothing should ever, replace the real" except, perhaps, that Harvey's fragile notions of what is "real" are somehow threatened by her understandings of the Internet. Most of all, I have no idea what it means to "unlearn how to be human," and I seriously doubt that Kerric Harvey has the metaphysical or spiritual transcendence to know either.

On the Internet, like elsewhere in contemporary society, issues of "reality" have become a slippery slope of polemics and vested interests. Debates concerning the "reality" of these kinds of experiences rapidly slide down the greasy and often dogmatic slopes of whose definition of what shall be considered "real" will be temporarily honored. We would be wise to think carefully about the assertions made by anyone that claims to know the ultimate truth on these matters. Knowing this, it may be prudent to suggest that questions about the "reality" of online experiences, communities, and forms of selfhood are closely related (if not identical) to the fundamentally unanswerable question of whether *any* experience, community, or self is genuine, veritable, true, or authentic. The difficulties of conclusively establishing the "reality" of online experiences, selves, and communities simply mirror and highlight the same ontological dilemma of any experience, self, or community.

As W. I. Thomas (1966) has pointed out, once defined as real, things become real in their consequences. "Reality" is more a matter of definition than anything inherent in self, society, or nature. Therefore, on the Internet or in everyday life, whether something *can* be experienced is much less important than the fact that people believe they do experience something; whether self *can* exist is far less important than the fact that people claim to experience it; whether a community *can* exist is far less important than the widespread belief among its members that it does exist. In short, when we define and perceive things as "real," we achieve something that is not only epistemological or ontological, but experiential as well (Aycock and Buchignani 1995). For these reasons, participants' claims to online experiences, self, and communities should be perceived as existing through the same processes of communication and interaction by which any other experience, self, or community emerges, is maintained, and is transformed in everyday life—the differences are to be found in the form by which they are experienced and the means by which they are produced, not in their degree of genuineness, truthfulness, or reality.

Self and the Means of Social Interaction

When people communicate on the Internet, they interact with others and create personally meaningful identities in electronic space (Baym 1995; MacKinnon 1995; Markham 1998; Myers 1987; Stone 1995; Turkle 1995; Waskul and Douglass 1997). This is neither surprising nor unusual; selfhood thrives everywhere people communicate and interact. Yet it is here where this analysis takes on a distinctive double character. On one hand, experiences of self and social situation on the Internet share much in common with experiences of self and social situation in everyday life. An online self is, at one level, no different from selfhood in any other context; a self on the Internet is something that is symbolic, communicated, presented, and negotiated. On the other hand, as Marshall McLuhan might insist, alterations in the means by which people communicate will subtly, yet powerfully, transform the boundaries and nature of social interaction and selfhood. When people interact by mediums of communication, they must translate themselves through the conventions of the media, and in that process, selfhood is necessarily transformed. Thus, the double character of this analysis: to understand how and why these experiences are different from those of everyday life and how and why they are similar, and thus to explore implications by peering into the interstices between these similarities and differences as people create new configurations of experience. The similarities between online and everyday experiences of personhood will be emphasized throughout this analysis. For now, it is necessary to frame our understandings of the differences.

In *No Sense of Place,* Joshua Meyrowitz (1985) details how electronic mediums of communication dislocate taken-for-granted relationships between physical setting and social situation, rearranging social experiences by overriding situational geographic boundaries and symbolic definitions. Meyrowitz's main point is that whether we are watching television, communicating on the telephone or by computer-networks, the social situation we are "in" is only marginally related to our physical location (if at all). For example, by use of cell phone, a person may be driving a car while enjoying intimate conversation with a phone-sex operator. There is no necessary relationship between the automobile and the erotic conversation; with electronic communication physical location and social situation are no longer necessarily connected. As our kinky commuter illustrates, an erotic situation no longer requires a bed, private space, or co-present corporeal bodies—it only requires voices, phones, and (in most cases) a credit card. Likewise, an online chat or cybersex participant may be at home or on an office computer, but he or she is part of a social situation that has little, if anything, to do with where he or she is located in geographic space. In this way, as electronic media dislodges social situations from physical settings, it rearranges taken-for-granted distinctions between them, blurs social spheres that were once distinct, and results in new configurations of social experience.

According to Kenneth Gergen (1991), these kinds of techno-social transformations have not only rearranged relationships between physical space and social situations, but also radically shifted our exposure and relationships with others, significantly altering everyday experiences of self and social world. By use of technologies of communication and transportation, we frequently encounter ever-increasing varieties of people from widely diverse social worlds, encompassing a boggling array of divergent viewpoints. The result is a state of "social saturation" where technologies make it possible to sustain relationships with others in an ever-expanding range of diversity.

As these technologies of social saturation (television, computer-networking technologies, transportation, radio, etc.) infiltrate the reality of everyday life, daily experiences occur in the context of an enormous range of other people, special interests, forms of relationships, unique circumstances, and opportunities for interaction—all of which alter the arena and materials by which selfhood is experienced. "With social saturation each of us comes to harbor a vast population of hidden potentials" (Gergen 1991: 71), where there is no individual essence or absolute standard to which one remains true or committed. Each "hidden potential" represents a possible self that under the right conditions may spring to life. In a state of social saturation, one's self is continuously emergent, reformed, and redirected as one floats along the torrid currents of a technologically sustained river of fluid and rapidly changing social relationships. In short, there is an ever-increasing possibility of developing a "populated self" or the acquisition of multiple potentials for being.

The experience of selfhood in online chat is a quintessential illustration of what Gergen (1991) means by a "populated" and "saturated" self. As almost everyone knows, the Internet provides a context for extreme self-fluidity; anyone can be anyone on the Internet. Sherry Turkle (1995) has written extensively on these experiences of self-fluidity, noting—as many scholars have before and since—that people create and cycle through a sometimes surprising range of anonymous online identities. Throughout her analysis, Turkle describes how computer-mediated forms of social interaction encourage people to play with a multiplicity of roles and experiences of self in a liminal threshold somewhere between what is "real" and what is virtual. "'This is more real than my real life,' says a character who turns out to be a man playing a woman who is pretending to be a man. In this game the self is constructed and the rules of social interaction are built, not received" (Turkle 1995: 10).[3]

As electronic media override taken-for-granted boundaries between social situations and physical space, we are instantly and efficiently taken to distant and divergent places in space and time. Electronic media present contrasting and conflicting images in a juxtaposed manner that blurs the modern sensibilities that have traditionally kept them apart. Through electronic media, one's self is exposed to a broad array of "we's" and "them's," each potentially representing what we were, are, and can be (Holstein and Gubrium

1994). Electronic media allow us to privately view public activities, publicly view private activities, develop distantly intimate relationships, perceive real events through a medium of representations, and perceive mediated representations as if they were real. The contradictory nature of these capabilities confounds taken-for-granted conceptions of reality, revealing a hyperreal dimension to electronic forms of communication.

Jean Baudrillard (1981) describes "hyperreality" as a simulation that is no longer of territory, referential being, or substance. The hyperreal is generated through an infinite Xerox of copies upon copies from an original that is no longer of known origin or veritable reality. It is a condition of pure simulation that threatens distinctions between "true" and "false," "real" and "imaginary." As Baudrillard (1981: 11–12) describes:

> it is no longer anything but a gigantic simulacrum: Not unreal, but a simulacrum, never again exchanging for what is real, but exchanging in itself, in an uninterrupted circuit without reference or circumference...it *plays at being* an appearance...it is no longer in the order of appearance at all, but of simulation.

Similar to other electronic mediums of communication, online interaction is a hyperreal simulacrum—neither real nor imaginary—a simulation that we may loosely refer to as "virtual." Online interaction is "real" insofar as people are really communicating and interacting with others. However, these interactions may best be described as a collective dramaturgy on an empty social stage (Baudrillard 1981), where the social production of space, place, and self are not contained, situated, dependent, or contingent on what is "physically real," nor are such productions necessarily grounded in that which is empirically verifiable. Online interaction proceeds as a simulacrum for which there is no ultimate referent, and whose status as "real" can only be assumed.

To summarize, cyberspace is a socially constructed hyperreal context of communication, mediated by technologies of social saturation that dislocate space, time, and personal characteristics of human interaction. Without referent or necessary commitment to the "physically real," online communications allow participants to construct new places, new social roles, and personally meaningful identities. With the click of a mouse, people can interact with diverse others in a multiplicity of socially produced places, providing a context for the experience of anonymous yet personally meaningful identities situated in a geographically spaceless context where commitments to any given self are as easily disposed of as they are accessed. The key question here, then, is how these fluid selves are constructed, how are they experienced, and what this form of selfhood can tell us about the nature of self in everyday life.

Cyberself: Selfhood in Online Chat

On the Internet, or anywhere else, self emerges in a process of communication and interaction. Important to this process are both symbolic meanings and the structure of interaction. That is, the meaning and form of interaction are symbiotically wed in communication, and this is especially significant for online chat since it is emergent *within* computer-mediated environments. For these reasons, we will begin with a brief description of the form of online communication—how participants communicate in online chat and the nature of those communications. We will then move toward an analysis of the symbolic systems utilized by online chat participants and illustrate the nature of "cyberselfhood."

The Multiple and Asynchronistic Form of Online Chat

Typically by use of client software, participants' logon and meet one another in chat "channels" (or "rooms"). Communication occurs in "real time"; in the here-and-now of the present, hindered only by the speed of participants' Internet connections, typing skills, and knowledge of the technology. For all practical purposes, it is a form of communication that is temporally similar to voice communications (like telephone) but mediated by text (like a written letter) within computer technologies (like e-mail). Because of the unique way that online chat blends these forms of communication, we might best describe it with hyphenated words like written-speech, print-conversation, or text-talk.

Online chat occurs in "chat rooms" or "channels" that are identified by a name. These channel names function to define both the context of communication and the content of the conversations that may be found within. Clearly, to go to a chat channel entitled "Christian Fellowship" is to know the context and content of communication that can be expected. Typically participants create these chat channels with a simple procedure, and the names of most chat channels are listed for all members within the system to view at any time (although some can be "private," or otherwise known only to a select few).

The names given to these channels are important outward cues to what people are chatting about and thus instrumental when choosing in which channel to participate. Indeed, as these labels reveal, there are a wide array of conversational themes from which to choose. At any given time, one may chat with others on matters of religion, politics, sports, various recreational activities, computer games, parenting, current events, and numerous other topics (just about any you can imagine). Sex and romance, as one might guess, are typically the most common subjects for online chat, and the abundance of channel names on these topics reflects this preference. "Hot chat" is not only the most common theme; sexual chat channels are also among the most frequently visited. Given the popularity of sexual themes, in an understandable attempt to control content many chat systems contain segregated regions specifically designated for teen chat, adult chat, computer gaming, news, etc.

Once a chat channel is chosen, participants may communicate with one or all others in the channel. Each person is identified by a self-selected screen name or "nick" (nickname), and usually, each chat channel contains a listing of all screen names "present." Hence, users always know whom they are communicating with in the channel (identified by screen name).

Chat communication is textual and synchronistic. All persons can immediately and simultaneously contribute to the conversation(s) by typing messages and clicking icons. Instantly, the message and the screen name of the sender will appear on the computer monitor of all other persons in the channel. Participants type what they have to say back and forth, and by this process they carry on lively synchronistic discussions. Persons in the channel also can simply watch written conversation(s) scroll up the computer screen, joining when (and if) they choose.

Because of its synchronistic nature and the absence of physical non-verbal communication cues, the linearity of communication in online chat is easily compromised. Conversations are frequently disjointed and will change rapidly. Thus, although synchronistic, online chat is often made asynchronistic by the multiplicity of conversations that occur simultaneously. Consider, for example, the following conversation, which came from a chat channel entitled "tarot" (the left-hand column identifies the screen name; the message follows the colons):

Trudygal	: what goes on in here? I am a virgin to all of this.
ASK4U	: you got cards savannah?
OSusanna	: I don't have to be a psychic to accurately tell fortunes, no one does
Washcycle	: that's what they all say Trudy
OSusanna	: It's just a matter of a good guess
DogsAime	: Is anyone available to do a reading??
Trudygal	: no really I am
SKYHOOK	: Oh Trudy, never tell that to anyone in here!
Washcycle	: There are official readings 9–12pm ET

Trudygal	: why??
Washcycle	: and just chat rest of time
Amulet	: hey trudy once you get the bill you wont be a virgin anymore!!!!
SKYHOOK	: I hear SysSage is reading, DogsAime
PVCDC	: who does the readings?
Trudygal	: I am already fearing that

As this brief excerpt illustrates, it is common for online chat to involve multiple simultaneous conversations. This is vastly different from the linearity of normative conversation that, not unlike a game of chess, is guided by implicit rules and maxims: only one person speaks at a time, signals are given, turns are taken, and the whole encounter is structured in cooperative activity (Wardhaugh 1985). While certainly cooperative, online chat departs considerably from this kind of normative sequentially structured conversation. Here, for example, "Trudygal" makes a statement that, three lines later, "Washcycle" responds to. On the seventh line "Trudygal" responds to "Washcycle's" comment. Then, on the eighth line, "SKYHOOK" responds to "Trudygal's" original statement. All the while, "OSusanna" carries on a conversation with various chat participants on matters of a different nature. And, all of this occurs in a few dizzying seconds. Such conversations ("chat") continue on and on, with each successive statement eliciting a response from any one or more persons within the channel.

These communications can initially be difficult to follow and can take a little getting used to. As one participant remarked, "conversations can easily become disjointed. It's often difficult to keep up with what's going on." However, like most anything else, it doesn't take long before participants adjust, learn how to interpret, and effectively participate in online chat. This isn't an especially difficult adjustment to make. After all, online chat is not to be read as an artifact—it is not to be read like a book or a magazine article. An excerpt like the one cited above falsely represents the nature of online chat. Online chat is a unique form of conversation; it is something that is subjectively experienced in a process of communication. Like other kinds of conversation, social and emotional meaning is conveyed through words but also through other non-discursive cues that contextualize those words. This is no less true for online chat. The meaning of these sometimes chaotic conversations is conveyed through words, but they are not to be read in a linear fashion, and even if they were, meaning is also importantly conveyed by the pauses, breaks, disjunctions, speed, and timing that are lost when these conversations are transposed into a text that is read out of the context of online chat. Furthermore, like any other group of people engaged in gregarious conversation, side discussions and secondary chitchat are common, and one deals with the variety of simultaneous conversation by only paying attention and responding to what is directed at him or her or what occurs within his or her immediate surrounding. Online chat conversations are similar to this kind of gregarious conversation and flow quite seamlessly for the same basic reasons.

While most communication within chat channels is available for all to view and participate in, there are other modes that allow for private one-on-one discussions. Private messages are communications sent from one user to another and do not appear in the "public" text of the channel; they are equivalent to whispering in someone's ear. Private messages can be sent and received by anyone in the chat system regardless of their online location. Private messages are so widely utilized that it is not unusual to enter a chat channel that contains numerous participants, yet nothing (or very little) is communicated within the channel—all communications occur through private messages. Furthermore, since there is no necessary limit to the number of distinct private messages that can be sent or received, they allow for tremendous flexibility in online communication and contribute even more to the asynchronistic nature of online communication. Participants can communicate with others who are not in the same chat channel, carry on numerous private conversations, and may do all of this while also participating in the "public" discussion that is occurring within the channel as well.

The important feature that must be emphasized about online chat at this point is its multiple and asynchronistic form. We will return to this form of communication as we seek to understand the nature of selfhood experienced by this means of communication. For now it is sufficient to recognize that online chat is a form of communication that entails a unique structure of multiplicity and simultaneous conversations among others who are similarly engaged in the same asynchronistic mode of communication.

Creativity and Constraint: Emoticons, Idioculture, and Online Chat

> Every group develops its own system of significant symbols which are held in common by its members and around which group activities are organized. Insofar as the members act toward and with reference to each other, they take each other's perspectives toward their own actions and thus interpret and assess that activity in communal terms. Group membership is thus symbolic, not a physical matter, and the symbols which arise during the life of the group are, in turn, internalized by the members and affect their individual acts.
>
> —Anselm Strauss, *George Herbert Mead on Social Psychology* (1964)

> Society not only continues to exist by transmission, by communication, but it may
> fairly be said to exist in transmission, in communication.
> —John Dewey, *Democracy and Education* (1916)

In online chat, communication occurs by typed text and typed text alone. Consequently, the obvious absence of physical presence, voice intonation, eye contact, posture, and other non-verbal cues poses certain constraints on chat participants. Most early scholars of computer-mediated communication were preoccupied with these and other ways that computer-mediated communication denies important communicative cues. Many of these scholars argued that the constraints of the medium hinder effective communication and result in predictable outcomes (Baron 1984; Cheseboro and Bonsall 1989; Kiesler, Siegel, and McGuire 1984; Rice 1984, 1989; Walther and Burgoon 1992). Because computer-mediated communication environments deny physical presence and related social cues, this approach to understanding and interpreting online communication and social interaction is sometimes called "cues filtered out."

When considering the suitability of the "cues filtered out" perspective to interpreting and understanding online communication environments, it is important that we remain attentive to two major factors. First, scholars of the "cues filtered out" perspective primarily derive supporting evidence from experimental research on task-oriented activities in organizational and occupational contexts. They were among the first scholars to become intrigued by computer-mediated communication and focused their attention on applications in organizational contexts where computer-mediated communications were first introduced. Second, scholars from the "cues filtered out" perspective generally seek to answer the question: How do computer-mediated communication technologies influence people, groups, and organizations? By examining task-groups in organizational contexts, these scholars identify elements of computer-mediated communication and causally link them to certain communicative outcomes. For example, participants in computer-mediated communications gain greater anonymity because elements of public identity are not immediately present (i.e. gender, race, age, and status). As a consequence, the nullification of differentiating personal qualities allows participation in task-groups to be more evenly balanced and egalitarian (Walther 1992), making it more difficult for some people to dominate or impose their views on others (Baron 1984). However, when everyone is allowed to express his or her views, it often takes longer to reach a decision or complete a task (Sproull and Kiesler 1991). Thus, computer-mediated communication heightens participation but impedes task resolution.

As both these factors illustrate, the "cues filtered out" perspective seeks to understand computer-mediated communications in a way that is largely one-dimensional and preoccupied by the question: How do these communication technologies influence communicative outcomes? What this kind of inquiry generally ignores, however, is the equally important question: How do people manipulate technology for their purposes? Or, as stated by Giuseppe Mantovani (1996: 93):

> The question: how do new technologies change people, groups and organizations? only deals with half of the problem. It should be completed with the second half: how do people, groups and organizations modify and adapt new technologies to suit them?

Although the "cues filtered out" perspective may be useful for understanding computer-mediated communications in work environments and in task-related activities, when considering conversational online communications (such as chat) it is clear that this does not provide an adequate model for interpretation or understanding. Perhaps the most important thing taken-for-granted by "cues filtered out" are the purposes for which participants use computer-mediated communication technologies and the conditions of their interactions. In computer-mediated communications, like any other kind of communication, the "task at hand" and the context of interaction influence the extent to which individuals are involved in what they say, do, what topics are raised, and how they are discussed. Task-oriented activities in work-related contexts represent one set of purposes and conditions for computer-mediated communication. Recreational contexts, like online chat, represent an entirely different set of purposes and conditions in which task-completion may have little bearing. Indeed, the only "task" in online chat is to communicate.

Understanding the role of "cues filtered out" in online chat might best proceed by perceiving the constraints imposed by the medium not simply as a hindering force, but as a structure by which participants focus creative and imaginative communication efforts. As suggested by Brenda Laurel (1993), it is possible that the constraints and limitations of the medium represent a structure that focuses creative efforts, and paradoxically increases imaginative powers by reducing the number of possibilities open to participants. That is, *heightened* creativity may arise out of the tension between the desire to communicate and the constraints of the medium, forcing spontaneity into various creative, imaginative, and inventive forms.

To begin with, there is little doubt that chat participants are acutely aware of the constraints of the medium. They understand how these constraints limit communication and pose certain problems to social interaction. As one chat participant said, "I dislike the fact that you can't communicate emotion in text. This often leads to problems with members reading something into the text that isn't there." In response to these conditions, "emoticons" were created. They represent one of the best illustrations of heightened creativity within the constraints and limitations of online chat environments.

Emoticons are a lively system of typed symbols invented and employed by chat participants as shorthand for expressing emotion and action. With emoticons, participants creatively construct and utilize text characters that symbolize socio-emotional, contextual, and interpersonal communication cues. These "emoticons" are symbols for smiles, frowns, laughing, pointing, and numerous other non-verbal cues for communication that, like Chinese calligraphy, are often literal representations of the ideas they convey (Marx 1994). While most people are probably familiar with these emoticons, here is a brief list of a few of the most common and their basic meanings:

:)	=	Amusement (notice the symbol is a sideways smiling face).
: (	=	Disappointment, sadness (frowning).
;)	=	Devilishness, playfulness, flirtation (winking).
{ }	=	Hug (place a screen name between the brackets to indicate hugging that person).
←	=	Pointing at one's self.

In addition to emoticons, chat participants create and employ a wide range of acronyms. These acronyms are shorthand for actions, reactions, and frequently used statements. Some of the more common acronyms used include:

LOL	=	"Laughing Out Loud."
LMAO	=	"Laughing My Ass Off"
BRB	=	"Be Right Back"
BBL	=	"Be Back Later"
WB	=	"Welcome Back"
JK	=	"Just Kidding"

These symbols and acronyms, in addition to hundreds of others (see Sanderson and Sanderson 1992), are used to indicate that a statement should be read humorously, sarcastically, in good spirits, in disappointment, surprise, and a wide range of other emotions.

Emoticons and acronyms are not the only mechanisms by which socio-emotional and interpersonal cues are communicated. Unconventional spellings, specialized grammar, and stylistic means for placing emphasis upon certain words are other mechanisms by which participants create a context for this kind of "written speech." For example, in online chat, sentences are often ended without periods, commas are often positioned to indicate pauses rather than clauses, and ellipses are used to indicate long thoughtful pauses. "Utterances" like "uh-huh," "yee-haw," "yippie," "hummm," "ooops," and "oh," are common. Uppercase letters indicate shouting ("SHUT UP! ...stop teasing me."). Asterisks are typically used to place emphasis on words ("I am *not* gonna talk about that!"). This agreed upon use of unconventional English, sprinkled with emoticons and acronyms, functions to heighten the sense of immediate spoken conversation.

Obviously, it is possible to write emotions and even emphases with words that indicate one is just kidding or being sarcastic, but doing so is not only time-consuming, it also undermines the directness and spontaneity of what is expressed (Marx 1994). Thus, emoticons, acronyms, and unconventional English are usually intertwined with text communication as a kind of gestural and emotional punctuation mark, as illustrated in the following dialogue:

SimpleGirl	: Good morning!!!
Shane	: hi SimpleGirl :)
Dan	: {{{{{SimpleGirl}}}}}
CmLaInT	: good morning Simple!
Brian	: uggh..vodka..no more drinky drinky for me the next uhmm...12..naah....13 hours :)
Ranger	: LOL brian!
Terry	: bye y'all, see you later :)
Dan	: bye terri ;)

Sometimes this kind of stylistic written-speech is used so expansively that it results in communications which would appear to outsiders as a mixed-up set of nonsense characters. In fact, when used extensively, online dialogues can become a form of communication that makes little more than hieroglyphic sense to those who are unfamiliar with the meanings of the symbols, as illustrated by the following:

Lucky	: Dabble !!
Dabble	: {{{{{{{{Lucky}}}}}}}}}
Dabble	: <-----Jon, fixing Dabbles pc

Lucky	: LOL
Lucky	: =: O
EndoEarth	: I'm back!!
EndoEarth	: <~~got booted
Lucky	: wb
EndoEarth	: thanks*******
Dabble	: <~~brb, downloading...
KittyKat	: Well all, it's been fun, but gotta do the work thing...Have a good one
KittyKat	: TTFN

Since words and keyboard characters are all that are available, participants creatively compress the richness of meaning into typed symbols that aid in the construction of emergent, fluid, and interpersonal communications—a form of "written-talk." In this way, emoticons, acronyms, and unconventional English are innovative forms of expression that vividly illustrate creative and spontaneous interpersonal communication in online chat.

Although the richness of face-to-face interaction cannot be entirely recaptured in the text medium of online chat, that richness is not entirely eliminated either. It is clear that in online chat, contextual, interpersonal, and emotional cues are creatively communicated and thus cannot be precluded as a meaningful element of online social interaction, as is sometimes suggested in the "cues filtered out" perspective. Indeed, there is ample evidence to suggest that the opposite is true. As detailed by Rollo May (1975), constraints or limitations represent a structure that channels creative energies, forcing people to manipulate available materials to construct innovative forms of expression. Online chat is limited to the characters on a keyboard. Yet, emergent from these constraints are creative forms of stylized interpersonal communication.

While the specific mechanisms by which participants creatively overcome the limitations of the computer-mediated environment represent interesting dimensions of online chat, the significance of emoticons and other contextual and interpersonal cues is not that participants "feel" emotions or "sense" a context. Rather, there are two more important issues to consider. First, emoticons, acronyms, and other creative cues allow participants a mechanism to communicate socio-emotional and contextual indicators to others and, hence, validate them in a social context. As others see and respond to these symbols, they become meaningful and experientially real.

Perhaps more importantly, one can readily identify "newbies" to online chat because they frequently ask about the meaning of these symbolic constructs, thus illustrating the extent to which they become a form of slang that marks insider status. In this way, emoticons, acronyms, and unconventional English serve an important second function; they are defining elements of an online *idioculture* (Fine 1987, 1979). "An idioculture consists of particular examples of behavior or communication that have symbolic meaning and significance for members of a group" (Fine 1987: 126). If culture is "a set of expressive and interpretive resources" (Marvin 1995: 1), then these online paralinguistic constructs become a cultural lexicon that identifies members of a distinct idiocultural group. Like occupational groups (McCarl 1986) and close-knit families (Zeitlin et. al. 1986), online chat participants have a specialized vocabulary based within their unique environment.

In general, idioculture can include a wide range of things, such as slang, clothing, jokes, nicknames, rules of conduct, hairstyles, music preferences, and so on. In any group culture, some elements of idioculture will have more significance than others, and, therefore, the most important are those that are repeated on many occasions (Fine 1987). Cues such as emoticons, acronyms, and the use of unconventional English represent one such repetitive and salient element of online idioculture. Because "knowledge and acceptance of a group's idioculture is a necessary and sufficient condition for distinguishing members of a group from nonmembers" (Fine 1987: 128), by the use of emoticons and other online communicative cues, participants display a cultural mastery and validate themselves as members of a larger "electropolis" (Reid 1991) of chat participants. Or, as Marvin (1995: 4) describes:

> Ironically, the use of these specialized symbols disrupts the illusion of virtual speech created by conventionalized misspellings and paralinqual smileys. They are the marks of "inside status" because they demonstrate knowledge and skill which are the requirements of belonging for a group with no kinship, geography or occupational ties.

To assume that constraints imposed by the computer-mediated environment hinder meaningful social interaction and cause particular forms of social organization, as often implied by scholars of the "cues filtered out" perspective, underestimates the resourcefulness of human communication and the nature of these creative social environments. In online chat, the constraints of the medium represent a structure that channels creative communication energies to overcome the limitations of the medium. An agreed upon language of emoticons, acronyms, and unconventional English represents the most salient of these creative measures. Moreover, these creative means become defining elements of the online environment in a cultural sense: an important element of an emergent and evolving online idioculture. Knowing and using these idiocultural constructs identifies and defines members of the group, distinguishing them as unique from others.

There is no doubt that "cues filtered out" poses constraints on communication; however, research on task-related activities in organizational contexts proves inadequate for interpreting the activities of chat participants. At the very least, task-oriented activities in organizational contexts involve a fundamentally different set of purposes and conditions by which people may use computer-mediated communication technologies, and these purposes exert an influence on what ultimately emerges from that form of interaction. When considering online chat, rather than being simply constrained by the medium, participants creatively exploit features of the online environment to develop new forms of communication and formulate new idiocultural constructs that are not derivative from the computer medium alone, but emergent from the interactions between persons in response to the constraints imposed by the medium. Clearly, one must consider the human participants and their purposes for interaction as having a critical role in the kinds of creative communicative outcomes that emerge in computer-mediated contexts—as we should for human communication and interaction in any context.

Cyberselfhood: Writing Self into Existence

> It is through various fictions that we endeavour to come to know ourselves.
> —Mike Featherstone and Roger Burrows,
> *Cyberspace, Cyberbodies, Cyberpunk* (1995)

Cyberselfhood begins with the creation of a screen name, which is of utmost importance in chat communication. In online chat, participants do not know who they are communicating with, but they do know which other screen names are present in the channel, and all communication is associated with the screen name of the message sender(s). Thus, the whole of a person's online presence must be condensed into a screen name—a single word or phrase. For these reasons, screen names are a critical element in the presentation of an online self; with screen names, chat participants convey important meanings about who and what they claim to be.

In Shakespeare's *Romeo and Juliet*, Juliet asks, "What is in a name?" She goes on to address the question metaphorically: "that which we call a rose would smell as sweet" even if it were not called a rose; Romeo would be every bit as wonderful even if his name were other than Romeo. Juliet's conclusion is a charming and pleasant thought. Unfortunately, Romeo's first name is not the problem—the problem is his family name, Montague. The Montagues are enemies of Juliet's family, the Capulets. Given what happens to Romeo and Juliet, it is apparent that there is much more to a name than Juliet cares to admit. We may conclude from this tragic story (and a mountain of sociological, psychological, anthropological, and linguistic evidence, which came long after Shakespeare's famous play) that language has an enormous impact on shaping meaning; neither Romeo nor a rose would be nearly so sweet if called by a different name.

Understanding selfhood in online chat requires that we, unlike Juliet, avoid this oversight, recognize the importance of names, and understand how they convey significant meanings that influence how we interact with others. Although in online chat a screen name may consist of just one or two words, as Haya Bechar-Israeli (1995) has illustrated, these words can evoke complex meanings and images—and this is not unique to the Internet. For example, we often choose familiar and comfortable names to neutralize frightening or alien entities (calling the Loch Ness monster "Nessy," or calling hurricanes by human names). A name may be used to glorify (calling a ship the "Queen Elizabeth"), or a name may be intentionally crafted to convey double meanings (thirteen-year-old boys, and some thirteen-year-olds who are much older, can endlessly amuse themselves by any mention of the name Mike Hunt or last name Johnson, especially when combined with a first name like Harry). Some names are so powerful that they become concepts in their own right (Marxism, Darwinism), often carrying ideological connotations (Reaganomics, Jacksonian politics) which can be evoked to emblematically represent an entire era (the McCarthy era). There can be no doubt that names are an important mechanism for conveying meaning, and they are no less important to the individuals for whom they are associated. As anyone with a nickname or anyone who has legally changed their name already knows, one's name is an important part of one's self and how others perceive one—something of which chat participants are acutely aware and actively manipulate.

As chat participants select, change, and manipulate various screen names, they begin a process of toying with their self in what will become a uniquely situated identity game. That is, by selecting a screen name, participants associate themselves with a self-selected label that is intended to convey crafted meanings. As chat participants quickly learn, their screen name is the prominent indicator of who they claim to be, their interests, and motives for interaction.

In many cases, screen names are simple derivatives of a first name. For example, a screen name such as "M1chele" is presumably based upon the first name Michelle. In some cases, this may be the chat participant's real first name; in other cases it probably isn't. In all cases, we cannot know for sure. After all, a first name will usually indicate a gender and, for various reasons, chat participants may intentionally conceal their gender, or present themselves as the opposite gender. First names are the easiest way to accomplish either objective. Some screen names indicate location. For example, "MrMaine" presumably indicates the geographic location of the person, although here too we cannot know for sure—perhaps "MrMaine" lives elsewhere but has some special interest in the state or some other reason for suggesting a personal association with Maine. Screen names may also be based on a person's hobbies and interests. For example, "GuitarPickn" would seem to indicate the person's interests in guitars and guitar playing. Other screen names are based on occupations ("TeachMan" is suggestive of a male teacher), while still more are based on life styles ("VegDiet" might indicate that this person is a vegetarian). Not surprisingly, there are a number of screen names that are based on motives for online interaction. Screen names like "PhoneFun4u," "M4Cyber," and "HotChic4Fun" unambiguously communicate sexual interests and motivations for online communication.

What is significant about these screen names is that they provide an important outward cue about who these people claim to be, how we might communicate with them, and what we might expect from those interactions. Chat participants are fully aware of this and actively manipulate these meanings to entice the kind of communications in which they are most interested. For example, if a chat participant were not interested in cybersex, they would be wise to avoid using a screen name like "SexyLady"; such a screen name would surely evoke innumerable invitations for hot chat. Likewise, if a person were interested in finding a partner for cybersex, they would be wise to approach people with screen names like "M4Fun" or "KinkyChick," while also avoiding chat participants with screen names like "Jesus4life." Regardless, screen names frequently indicate how we may communicate with online chat participants. If a person were interested in chatting with "Jesus4life," he or she would already know where to start a conversation—the screen name is a cue to inquire on matters of faith.

In everyday life, clothing, hairstyles, accessories, and a huge assortment of various other material and contextual cues convey symbolic information about the kind of people we are. Throughout his work, Erving Goffman magnified the importance of these kinds of symbolic cues, brilliantly analyzing how they are used in a twin process by which people both "give" and "give off" crafted impressions of self. Because these cues are not immediately present in online chat, screen names function as a kind of substitute. Without physical presence, screen names become the only initial means by which chat participants can communicate qualities of selfhood that are normally observed (gender, age, geographic location), discerned by social cues (marital status, social status), or acquired through knowledge of the person (hobbies, interests). In this way, screen names are important components of the presentation of an online self; they are "trademarks, distinctive individual smells by which their users are recognized" (Myers 1987: 240).

As important as screen names are, they are just one facet of the presentation of an online self. In the final analysis, a self does not reside solely in a name but emerges in communication and interaction. In online chat, or anywhere else, a self must be presented, negotiated, and validated in a situated context among others. This process can be routinely observed in almost any chatroom at almost any time; it goes on perpetually. For example, consider the following:

RedWines	: Let's have an age/sex check
LeFetes	: 24/F
EdsFerret	: <-------26 F
RedWines	: 20/f
DocNut	: 27/m
RedWines	: How about a state check?
RedWines	: MD
EdsFerret	: OH
DocNut	: MI
LeFetes	: FL
EdsFerret	: What is your occupation Doctor???
DocNut	: psychologist....YIKES!!!
EdsFerret	: Oh, no I already feel like I'm being analyzed :)
DocNut	: where ya from EdsFerret??
EdsFerret	: Toledo, Ohio
LeFetes	: Hey RedWines, what do you do?
RedWines	: I work with retarded adults
RedWines	: and what do you do?
LeFetes	: I'm a secretary?
LeFetes	: ...and I play the guitar
EdsFerret	: Hello, LeFetes. is that a French name?
LeFetes	: Actually I'm Italian
LeFetes	: ...But my moms French
Ball 0	: Hi everyone!
EdsFerret	: That's neat, do you know French?
LeFetes	: me oui moin ami

LeFetes	: yes
EdsFerret	: moi aussi aujourd hui.
Ball 0	: Any college students here?
AndyCapps	: Yes, Miami Univ. in OH
EdsFerret	: The University of Toledo
Ball 0	: Univ. CA, San Diego
Ball 0	: What are you studying, Andy?
AndyCapps	: Accounting. and you?
Ball 0	: Psych
Ball 0	: I'm learning to mess with people's minds. :)

This brief conversation is characteristic of online chat. Conversations typically begin and continue (on and on) as illustrated above. Because chat environments deny physical presence, nothing can be directly observed—everything must be communicated; all elements of selfhood must be presented to others; one must literally write one's self into existence. This process is not especially difficult; in fact, it seems to happen spontaneously. Once participants have stated basic information about themselves (usually age and gender), conversations quickly develop in relationship to that information and participants reveal more about who they claim to be—often with very playful overtures (also illustrated above).

What are explicit in this process are the *self-claims* that are made. That is, participants claim information that pertains to who they are. For example, even in the short excerpt cited above, self-claims proliferate and a considerable amount of information is conveyed. We have learned, for instance, that LeFetes is a twenty-four year-old female from Florida. She is a secretary and plays guitar. She considers herself an Italian, but her mother is French. She is (at least to some degree) bilingual, able to communicate in English and at least some French. What is important about these self-claims is not that people make them, but rather, that they have no choice but to make them if they wish to participate in online chat to any meaningful degree. In this way, online chat is organized around what is perhaps the most basic tenet of human interaction.

While there is nothing in online chat that requires it, everything about human interaction is linked to a simple rule: We cannot interact, communicate, or even be in the presence of others and fail to communicate something about who we are. So basic is this rule that in everyday life, the Internet, or anywhere else it is most often taken for granted and hardly noticed. Yet, this rule is absolutely central to the experience of life in society, so much so that it represents the foundation of Erving Goffman's (1959:1) insightful analysis, which begins:

When an individual enters the presence of others, they commonly seek to acquire information about him or to bring to play information about him already possessed. They will be interested in his general socio-economic status, his conception of self, his attitude toward them, his competence, his trustworthiness, etc. Although some of this information seems to be sought almost as an end in itself, there are usually quite practical reasons for acquiring it. Information about the individual helps to define the situation, enabling others to know in advance what he will expect of them and what they may expect of him. Informed in these ways, the others will know how best to act in order to call forth a desired response from him.

Goffman goes on to explore the various "sign vehicles" by which this information is conveyed, and how this "sets the stage for a kind of information game—a potentially infinite cycle of concealment, discovery, false revelation, and rediscovery" (1959:8)—not at all unlike online chat. Thus, on the Internet, like anywhere else, people who are in the presence of others acquire information about each other. Self-claims in online chat are not a departure from, but rather an extension of, this basic principle. In online chat, this information is acquired through screen names, but more importantly, through deliberate self-claims that participants communicate to others in text. The difference is that, due to the absence of physical presence and contextual cues, the self-claims of online chat are much more explicit, blatant, and obvious than the kind of subtle symbolic craftiness by which these claims are made in everyday life. Another important difference comes from the fact that self-claims in online chat are much more difficult to verify; the "rules" of the "information game" to which Goffman (1959: 8) refers are ambiguous and ephemeral. For example, because "she" is not physically present, we cannot know for sure if LeFetes is really a woman, twenty-four years old, Italian, or anything else "she" has claimed, although we cannot always be any more sure of these things in everyday face-to-face interaction either. Thus, other chat participants may accept, challenge, or simply ignore these claims but, generally speaking, unless there are gross discontinuities, self-claims are rarely contested to any significant degree—just as they are rarely contested in everyday life. Instead, when chat participants make self-claims, additional information may be asked and given, inquiries may be made, details may be provided, and in this process a cyberself is interactively crafted through the course of communication.

In crafting a screen name and interacting with others in chat channels, a "cyberself" is established. As already emphasized, cyberselfhood shares all the essential characteristics of any other self: it is something that is symbolic, presented, negotiated, and validated in a process of communication and interaction with others. Which is to say, cyberselfhood is emergent. As one participant said: "I didn't think through the creation of identity, it just evolved. I just make it up as I go along, as I talk to people. I don't change online; I just evolve." Furthermore, like any other self, a cyberself is situated; in this case, it is situated squarely in a computer-mediated communication environment. Thus, a cyberself may be defined as an emergent set of transient meanings that are crafted by individuals in association with others that temporarily refer to a person who is situated in the ether of electronic space. Or, stated differently, a cyberself refers to the meaning of personhood or experience of personal identity emergent within communication technologies and situated in interactions between people in these contexts. Like any other self, a cyberself is presented and negotiated in an ongoing process of communication. As one interacts with others, a cyberself is momentarily validated. Cyberselves are always situated performances that exist at the surface of a communicated present. To state it bluntly, a cyberself is always whatever is passing for a self at the moment in an electronic computer-mediated context.[4]

Anonymity and Disembodiment: Hyperfluidity and Doubt

Anonymity is an important part of online communication and interaction. As we have already learned, in the anonymous environment of online chat, presentation of self is everything; with screen names and self-claims, anyone can present themselves as anyone. To compromise anonymity is to compromise the supreme power to construct any self that one desires at any moment. Thus, as many participants have indicated, anonymity is highly valued because it is the means by which dynamic and fluid cyberselves are constructed:

> Online allows much more freedom—it's anonymous—whatever mood you're in you act upon. We all project an image with others—and are pretty consistent in person—online it doesn't have to be consistent.

> The anonymity factor can be intriguing, it gives one the freedom to chose which room to go into, whom to talk to and when, what to say, to lie or be honest while also knowing that every other user has the same freedom.

> It [anonymity] gives me the opportunity to express myself with no one to stare or wonder and not worry about where I was or with what company I was keeping.

As these statements illustrate, the anonymity of online chat allows participants the option to be someone different. So long as no one knows who you are, it is not necessary to be yourself. As one chat participant explained, "I'm constantly changing my alter-egos and identities on this silly system. You think I could get a life or somethin'." However, online chat pushes the envelope of this self-fluidity even further. Online chat is not only anonymous, but also a uniquely disembodied form of real-time communication. Sitting at a computer keyboard typing messages back and forth, selves are communicated but bodies are nowhere to be seen. With the corporeal body hidden behind the scene of interaction, participants are free to be without being; online chat dislocates selfhood from the constraints of the body; self is not contained or affixed to any physical form. Thus, it is anonymity that allows participants to be someone different, but it is disembodiment that transforms all categories of personhood into possibilities for the self that can be instantly realized, altered, and even deleted. For this reason, in the anonymous and disembodied environment of online chat, selves are not only fluid, they are *hyperfluid*. Properties of selfhood are transformed from nouns to verbs—self-selected, fluid, and dynamic identity utensils that are more or less useful for the purposes of communication and social interaction. All categories of personhood become pure labels: symbols to think and interact with, not within. Many respondents report this to be a liberating experience of free self-expression:

> Sometimes I pretend I'm a woman, I've also invented experiences. It enables me to play out fantasies. It allows me to take dreams one step closer to reality.

> You can be anything. I may stretch truth and be with whoever I want—no inhibitions.

> Online can be fantasy—You can be anyone you want to be. It also allows some to exercise a fantasy existence that they would never dare in real life.

Although many participants value the freedom of hyperfluidity, these conditions pose certain problems to cantankerous issues of legitimacy. To illustrate, consider the Net folk story of Douglas Adams. According to Net folklore, Adams, having acquired celebrity status for his wildly popular *Hitchhiker Guide to the Galaxy,* paid a visit to his own online fan club. However, upon arrival he was ruthlessly accused of being an impostor, and no amount of biographical trivia or argument could sway the consensus that he was a fraud. Consequently, Adams was not allowed to participate in his own electronic collectivity of enthusiasts. This Net folk story (of unknown authenticity) illustrates one of many ways that the "freedom to be anything" can become a source of frustration and concern. To further illustrate, consider the following responses to the question "What, if anything, do you dislike about online chat?"

> I don't know if people are as they claim to be. Many people are not what they claim to be. Men posing as women, etc!

> I enjoy the opportunity to meet people anonymously, but people often misrepresent themselves and turn out to be vastly different than their profile and behavior online would indicate.

> Because people can take any identity they want, they do. People who are 50 tell you they are 25, males tell you they're females, etc. People lie about who they are.

> There are a lot of very very horny guys out there, and so much so that they sometimes pretend they are women to get the bi or lesbian girls to talk to them. I don't know if that's a sexual game they play for their own fantasies, or if they're just lying assholes.

These problems are perennial to online chat; they are endemic to anonymous and disembodied forms of communication and interaction. In everyday life, someone who is not what he or she appears to be occasionally fools us. But on the Internet the art of the con is reduced to such relatively simple techniques that most people can reasonably manage to use them, and we are all made into potential fools. Looking at it pragmatically, on the Internet an "adult" is anyone who clicks a button that says, "Yes, I am over 18"; a "woman" is anyone with a female screen name; an "African" is anyone who says so; and so on. Each presents potential problems of authenticity that make anonymity and hyperfluidity both a value and a source of frustration.

In spite of these problems, interaction persists through what Erving Goffman (1959: 9–10) calls a "working consensus." That is, "together the participants contribute to a single overall definition of the situation which involves not so much a real agreement as to what exists but rather a real agreement as to whose claims concerning what issues will be temporarily honored." As MacKinnon (1995) notes, because participants are denied direct knowledge of one another, they must suspend or "forget" about the person behind the persona and rely on the individual's word as an accurate representation of self. In fact, as Douglas Adams learned, the ultimate challenge to online interaction is the possibility that some participants will *not* suspend doubt. In this way, there is a certain epistemology at work in online chat, and failure to interact with others on the basis of that epistemology may be the most disruptive thing a chat participant can do. This epistemology is founded on the obvious fact that in computer-mediated environments, representations are all that exist, and since there is no way to conclusively determine if representations adhere to what is true or real in the off-line world—all presentations of self are potentially suspicious—there is no way of sifting out the "real" from the "fake." Thus, in order for interaction to proceed, it is necessary for participants to suspend doubt and formulate a "working consensus" which identifies not so much what is "real" but what claims concerning which issues will be temporarily honored. Several participants noted this process:

Question	: *If you don't actually meet the person, how do you know if they are what they claim to be?*
Response	: You don't! That's part of the appeal! Some u meet some u don't but u get right to the heart of an issue without getting caught up. If u don't meet what difference does it make? As long as you have a good time, who cares!
Question	: *How can you tell if the people you meet are "genuine" or "fake"?*
Response	: I have yet to be disappointed. I just go on instincts, trust, and faith. Mostly I go on how they treat me and what I pick up along the way.

By agreeing on a working consensus, whether aware of this agreement or not, chat participants initiate a form of engagement—a willing suspension of disbelief (Laurel 1993). Chat participants generally understand that cyberselves are dramatic enactments. Or, in other words, participants are acutely aware that in online chat people enact self-performances that, like any other dramatic performance, are not necessarily real. However, to enjoy the drama (the chat), one must at least temporarily suspend knowledge that it's all "pretend"—hence the dramaturgical epistemology of online chat. This suspension of disbelief affords users a certain privilege of engagement where *"pretending that the action is real* affords us the thrill [of the drama], *knowing the action is pretend* saves us from the pain of [reality]" (Laurel 1993: 113, emphasis in original). As Brenda Laurel (1993: 115) explains:

Engagement is what happens when we are able to give ourselves over to a representational action, comfortably and unambiguously. It involves a kind of complicity. We agree to think and feel in terms of both the content and conventions of a mimetic context. In return, we gain a plethora of new possibilities for action and a kind of emotional guarantee. One reason why people are amenable to constraints is the desire to gain these benefits.

Through engagement, participants can experience the representational and dramatic world of online chat directly without disruption or distraction. This requires that participants proceed with interaction under the assumption that what you see is what you get, and the representation is all there is. As described by Annette Markham (1998: 26), "just as I don't assume everyone online is really what or who they say they are, I realize most people probably don't believe (or care) that I'm really what or who I say I am, either. For me, this provides a comfortable, protective façade." Thus, agency is bestowed unto online self-agents in a process that involves a willing suspension of disbelief and engagement in the drama of self-production.

Communication Play: Self-Games and the Multiplicity of Cyberselves

Not only can online chat participants present one cyberself, they can also have multiple screen names—each independent of the others—providing an opportunity to construct multiple anonymous cyberselves. As several chat participants explained, multiple screen names allow for multiple cyberselves that are often divided according to various motives for interaction and defining characteristics of selfhood:

> One name…is known only to my closest online friends. I use that name when I don't want to be bothered by strangers. The other names are used when I go in to chat rooms or private rooms where I wouldn't normally want my friends to find me.

> I actively use four of my five screen names…each screen name represents a different persona.

> (Q)*How many screen names do you currently use?* (A) 3
> (Q)*Explain why you use more than one screen name*: (A) I am curious to see other people's different reactions to various "personalities."

This multiplicity of anonymous self-enactments often results in communications that may be metaphorically called an ongoing "self-game." That is, online chat participants often actively and consciously play with multiple representations of who they are in these liminal online environments. After all, online interaction is a pastime that is usually done for the purposes of leisure and enjoyment, and thus, we should not be surprised that online interaction and cyberself enactments often become a literal form of communication play. As some chat participants explained:

Some people take this as a total game—I call it Nintendo for adults!!

I'm sure people modify who they are online, it's part of the game I guess, the mystery and fun of discovery.

I enjoy talking to people and having fun without being tied to "who we are."

It's quite fun to create a completely new identity. I have three others!

I have a different persona for each moniker. One is very sexy, but I have one who is grouchy and a real bitch. Another sarcastic, and another is very sweet. I know it sounds very schizophrenic, but actually, it is a lot of fun.

As these responses suggest, for many participants online chat is a communication game that is all about "playing with yourself" (in a literal rather than figurative sense). As one chat participant appropriately explained, "Like anything else, online is a game...isn't everything a game? Only here, I play with who I am." Furthermore, in the self-games of online chat, there is little commitment to any given self. In online chat, one's self is a fluid horizon of possibilities, and for many participants that is precisely what makes it fun. Each user has the power to create an unlimited self through interaction with others, and toying with these immense possibilities is what the game is all about. Several chat participants indicated this when they said:

It's kinda like split personalities. Shows that someone is only who they want to be when it's most comfortable.

I use different screen names to fit the environment of the chat channel.

I think if a person has a good imagination the skys the limit—you can become anything you want!

As these respondents clearly indicate, online chat often acquires game-like qualities. If, however, we are to adopt a game metaphor for understanding selfhood in online chat, it is important to qualify this with three additional insights. First, "play" and "game" are used as metaphors to highlight an often perceived playfulness that chat participants attribute to their interactions and presentations of self. Furthermore, to interact with others in online chat environments is to translate oneself into the conventions of the medium, and these conventions are like rules to a game that participants often "play" with. However, playfulness does not imply that all is fun and games. Persons can (and do) form deeply meaningful interpersonal relationships in the course of playful encounters. Playfulness does not eliminate capacities for such things as commitment and trust but merely makes such qualities playable, like any other quality of human interaction (Kristeva 1987). Second, if we are to adopt a game metaphor for interpreting online leisure-social activities, it is important to recognize the distinctions between "finite" and "infinite" games (Carse 1986). A finite game is played for the purpose of winning; an infinite game is for the purpose of continual play. Finite games have distinct rules; in infinite games the "rules" change during the course of play. Players of finite games play within boundaries; players of infinite games play with boundaries. Clearly, if a game metaphor is to be adopted, the metaphorical game is infinite. Third, we would be wise to avoid a "play" verses "serious" dichotomy—it is a false binary distinction that does not hold empirical clout. Serious play is a prevailing standard of any gaming activity, and just because some people are "playing" does not mean they may not take the activity seriously. Like all kinds of games, much online interaction can be found to be playfully serious.

Ultimately, it makes no difference whether a person's intentions are genuine or non-genuine, or whether interactions are playful or serious. In the final analysis, the meaning of the enterprise is established in the expressive-impressive dimension of communication (Brissett and Edgley 1990). Therefore, instead of embarking on attempts to impose binary play versus serious or real versus fake distinctions on the processes of online interaction, analytical attention should focus on the production of meaning—a process that remains consistent across perceived categories of "play," "seriousness," "genuine," and "fake."

In summary, when online, just as in face-to-face situations, a self emerges through interaction with others. The structure of online chat is multiple and simultaneous. Not surprisingly, so too is cyberselfhood. Parallel to the multiple and simultaneous channels of online communication exist a multiplicity of cyberselves. Each cyberself is an anonymous set of meanings associated with a screen name that may be presented as virtually anything. These interactions become a form of dramatic communication play—a hyperreal simulacra of communication, and a simulacrum of the self.

Conclusions

Online chat is a unique form of communication and interaction situated in a computer-mediated environment that alters the significance of time, space, physical location, and the corporeal body as interaction variables. It is the *form* of interaction that transforms social spheres into new online environments, with new patterns of social interaction. Furthermore, it is the form of online chat that allows for a fluid multiplicity of cyberselves that may be realized or deleted at any moment.

The form of computer-mediated communication can be described by numerous different qualities (see Waskul and Douglass 1997). However, online chat reveals two quintessential characteristics that are essential to understanding this form of communication and interaction: it is uniquely *dislocated* and *disembodied*. Online chat and most other forms of computer-mediated communication occur in a "place" without geographic "space"; in a real-time communicative environment without corporeal co-presence. Dislocated from the constraints of geographic location and the corporeal body, online chat is an ultimate context for self-multiplicity—a place for the experience of a no-holds-barred orgy of extreme potential self-fluidity. This character of computer-mediated communication poses a unique range of communicative possibilities for selfhood and social interaction. For example, once again consider the following situation briefly mentioned in Chapter One:

Wife Accused of Cybersex Cheating
Sommerville, NJ –A man filing for divorce accused his wife of carrying on a "virtual" affair via computer with a cybersex partner who called himself "the Weasel."
Diane Goydan's relationship with the man apparently never was consummated, but her husband, John Goydan of Bridgewater, claimed the pair had planned a real tryst this weekend at a New Hampshire bed and breakfast.
Goydan filed divorce papers Jan. 23 that included dozens of e-mail exchanges between his wife and a married man she met on America Online. The man, whose online name was "The Weasel," was identified in court papers only as Ray from North Carolina. (Associated Press, Feb. 2, 1996)

This highly publicized case of "virtual infidelity" succinctly illustrates how and why disembodied communications in dislocated Internet social environments depart from interactions in everyday life. After all, in everyday life, the political and legal status of "married" is, and always has been, firmly affixed to the corporeal body. That is, regardless of the range of roles and self-enactments in which one engages in everyday life, the status of "married" is firmly affixed to the physical body (emblemized by physical symbols affixed to the body, namely a wedding ring on a specific finger). The body represents the tangible thing to which a person is acknowledged as a "citizen," and is thus bestowed with the rights, duties, and responsibilities of any political or legal status. No matter where that body shall go or what the person does, the status of "married" legally and politically applies to the body.

However, in a disembodied context, participants can enact performances that are freed from these constraints, thus posing a unique range of interactive possibilities that challenge social, cultural, legal, and political conceptions of self and social interaction. Another more recent case further illustrates the nature of these problems:

E-Mail to Girls Leads to Man's Arrest

Waukesha—A federal government employee and former police dispatcher from Iowa drove nine hours to Brookfield to meet twin 12-year-old girls he met over the Internet and planned to molest, marry or adopt, according to charges filed Monday. Timothy Peter Koenck, a soil conservation technician for the U.S. Department of Agriculture in Iowa, was arrested about noon Friday at a McDonald's restaurant near Brookfield Square Mall, where he arranged to meet the two girls. But the girls turned out to be special agents with the state Department of Justice, according to a criminal complaint charging Koenck with *two* counts of felony attempted child enticement. If convicted Keonck, 40, would face a maximum 60-year prison term…According to the criminal complaint and other documents:

*Wisconsin Department of Justice officials were contacted after a private Internet watchdog organization called InternetWatch reported inappropriate e-mail from a user on Yahoo! Web sites.

*A woman with the group told authorities she posed as twin 12-year-old girls named Teddie and Georgie and received sexually explicit solicitations from someone using the computer name "dirtyboy69." Yahoo! linked that user name to Koenck.

*Koenck had e-mailed the fictitious girls about 20 times since December. He allegedly said he was interested in marrying one of the girls and having sexual intercourse or contact with both…Koenck then arranged to meet them, with Wisconsin Department of Justice agents now posing as the twins over the Internet. (*Journal Sentinel*, July 10, 2000, emphasis added)

If an adult has text cybersex (typed words, no physical contact) with a minor, most people would consider the act immoral and possibly criminal. The situation, however, is complicated. If, by having cybersex with a minor, an adult is guilty of a crime, then is an adult guilty of the same crime if he or she has cybersex with another adult who claims to be a minor? Would an adult be guilty of a crime if he or she has cybersex with a minor who claims to be an adult? Would a minor claiming to be an adult be guilty of a crime if he or she has cybersex with another minor? These situations are hazy at best. Timothy Koenck's apparent crime, however, is not hazy—he arranged to meet two fictitious minors for the purposes of sexual relations, both of whom were actually adult agents of the Wisconsin Department of Justice. Koenck's apparent criminal intent was not unambiguous. But, even so, *two* counts of child endangerment?

Situations like these raise new questions; they magnify the extent to which disembodied and dislocated forms of communication depart from other forms of interaction, sometimes resulting in controversial situations that are problematic to traditional understandings of selfhood and its relationship to society. Also, by posing a new range of possibilities that depart from traditionally conceived understandings of self and social world, new problems arise. Clearly, widely publicized cases like "virtual adultery," the "virtual cross-dressing psychiatrist" (Van Gelder 1985), raise serious questions about the nature of personhood and social interaction on the Internet.

In what ways, then, does selfhood on the Internet challenge taken-for-granted relationships between self and society? How do online chat participants negotiate these challenges? Is interaction on the Internet truly disembodied, or does the body remain an important component of selfhood, albeit transformed into something different? Most importantly, what do these experiences tell us about the nature of personhood in contemporary society? These will be key organizing questions for the remainder of this book.

NOTES

1. Although this is a shamelessly simplified explanation, Marx argued that the "material base" or "substructure" of society is the driving power of history, the nucleus of social and cultural change, and organizing force of society. This "substructure" is, to Marx, comprised of two critical elements. The first are "forces of production" of which Marx means a combination of labor, natural resources, and technology. The second are "relations of production" which refer to the relationships among people and symbolic systems of meaning that emerge as people make use of that labor, natural resources, and technology. For our purposes, it is sufficient to acknowledge one small component of Marx's remarkable insight: he not only recognized relationships between technology (forces of production) and the symbolic universe of human social and cultural activity (relations of production), he also magnified these relationships as the central, dominant, and organizing force of society and history.

2. "When Coca-Cola Company urges us to buy its soda over someone else's because Coke is the 'Real thing,' just about anyone can see how the term [real/reality] is being used in an effort to regulate and influence our patterns of consumption. But when we are told that America has a 'drug problem,' and furthermore that this problem is 'real,' it is not always as obvious that someone is trying to get us to buy into their definition of what is important" (Brissett and Edgley 1990: 35). This very same observation can be made regarding the precarious endeavor to define experiences of online social environments as somehow "less real" than other experiences of life in society.

3. As strange as this quote might read to someone unsocialized to these Internet communication environments, circumstances like this are not uncommon. For example, I played a character on a fantasy adventure MUD for a little over a year. It was a MUD based

loosely on a kind of *Dungeons and Dragons* fantasy role-playing. Characters are developed as they battle extraordinary beasts, obtain treasure, and advance in skill and ability. Like *Dungeons and Dragons,* participants in the MUD battle alone, but most often together, and sometimes against one another. I played a character that, with significant aid from other more established players, grew to be brutal and dangerous. Armed to the teeth and skilled in the use of those arms, he was a serious threat to those who found themselves on the wrong side of his sword (although he was by no means the most powerful on the MUD). Perhaps as a joke, or possibly to make a point out of my character's overt machismo violence, one of the deities on the MUD (a player that is all powerful, and typically one who oversees the more mundane technical side of day-to-day MUD management) turned my character into a woman. Thus, much like this person quoted from Turkle's study, I could say that I was a man playing a woman who was once a man. Complicating matters even further, the deity eventually turned my character back into a man. Thus, I was a man playing a woman who was once a man who is now a man again. Circumstances like these, as strange as they might seem, are not at all unusual.

4. I should point out that frequent visitors to the same chat channels who communicate with the same people over an extended period of time come to be known to others. In this way, cyberselves can develop both a history and some degree of consistency. Several chat participants indicated this when they said:

    I use this screen name because people know me by it.

    I prefer to go to the chat channels where people know my screen name.

    I use one name most of the time and it has the most recognition as "me."

    I use this screen name because it's well known by the people I chat with.

    This, however, does not negate the transitory nature of cyberselfhood. Rather, this highlights the same processes by which any meaningful relationship or self-performance emerges and extends beyond the surface of a communicated present. Furthermore, this highlights the processes by which meaningful personal relationships can—and do— emerge in communications on the Internet. Quite simply, as all selves are situated, emergent, and transitory, so too are cyberselves; they are situated, emergent, and transitory in a computer-mediated context. Yet, like any other self, when persons consistently interact with a relatively stable group of others through time, non-temporal roles sediment. The meaning of these roles draw referent from both the history of an individual's interactions and communications with others and broader socio-cultural structures of self-symbolism (i.e. people formulate conceptions of what "kind" of person he/she is on the basis of information provided)—cyberselves are no different.

❖ CHAPTER THREE

Accounting for the Cracks: Self-Multiplicity and the Cultural Prerogative of a Unitary Self

Specific procedures of universe-maintenance become necessary when the symbolic universe has become a problem. As long as this is not the case, the symbolic universe is self-maintaining, that is, self-legitimating.

—Peter Berger and Thomas Luckmann,
The Social Construction of Reality (1966)

...true self. No other concept so dominates the literature of despair and pessimism in our society; no other prize is so frequently promised in the guides and handbooks to alternative reality. The link between this goal and the pluralization of our life worlds is evident. We see our existence as made up of different life-worlds, various phenomenal universes within which we display desperate modes of consciousness. Our very lack of full commitment to any or all of these worlds produces within us the sense of some entity which stands back from reality, an entity which is presented within all of them but which is fully realized in none.... Our new search for a real self may have done little more than add another self conception to the stock derived from our supposedly "unreal" involvement in work, marriage, bureaucracy, leisure.

—Stanley Cohen and Laurie Taylor, *Escape Attempts* (1992)

While the Internet remains a new and emergent field for social thinking, there is sufficient research to warrant a few tentative generalizations. When considering anonymous online leisure communications—like online chat—three generalizations seem appropriate. First, participants construct self-performances in a dislocated and disembodied context that allows the freedom to "be" without "being." Second, online chat is, in and of itself, a multiple, simultaneous, and asynchronistic mode of communication. Third, within this disembodied, multiple, simultaneous, and asynchronistic communication environment, participants can enact selves that are equally multiple, simultaneous, and fluid. In short, online chat reveals a kind of infinite dramaturgical self-game, situated in a structure of multiplicity and dramatic self-fluidity that is frequently realized as some participants play with who they claim to be, and all participants know they may do the same.

Although situationally variable, these elements of online chat seem self-evident.[1] If we are to agree with this general characterization of anonymous online chat (similar accounts may be found in Featherstone and Burrows 1995; S. Jones 1995, 1998; Markham 1998; Porter 1997; Reid 1991, 1994; Turkle 1984, 1995), then serious consideration ought to be given to the relationship between these conditions and the more general cultural prerogatives that are said to "shape" and "mold" the "formless" "vacuum" (Geertz 1965) of the experience of life in society. In considering this relationship, one disjuncture is inescapably obvious: the experiences of self-multiplicity and self-fluidity that characterize online chat conflict with deep-seated cultural prerogatives for a paramount unitary self.

In the highly individualistic culture of contemporary western society, individuals are socialized to have a unique, singular, unitary, paramount self. That is, persons are socialized to have *a* personality, *a* core self, *a* relatively stable set of personal characteristics that are carried with them from place to place, situation to situation, as a constant referent to whom they innately are. "The modern self-concept defines each of us as an individual, with a distinct identity (from the Latin *idens,* meaning the same) that remains the same wherever one goes" (Anderson 1997: xiii). Kenneth Gergen (1994) appropriately describes this cultural prerogative as the belief in the "self-contained individual." Although people experience occurrences of self-disorganization and various crises of self-validation, nonetheless, the attempt to cultivate a "true self" remains a powerful cultural ideology that, in spite of the potential for personal crisis, continues to organize and provide meaning for the experience of life in society. Thus it would seem that the multiple and fluid forms of selfhood that appear so self-evident in online chat potentially challenge this cultural ideology and related experiences of self and social reality.

Given these conditions, how chat participants conceive, frame, and negotiate this apparent conflict is important to understanding at least one aspect of the relationship and implications of online chat to the general experience of contemporary selfhood. This chapter examines precisely this issue: How do online chat participants account for the obvious conditions of self-multiplicity in lieu of the cultural prerogative of a unitary self? This chapter will also attempt to address a related and equally important question: What do these negotiations indicate about the experience and cultural mechanisms for appropriating self in the contemporary world?

A Cultural Crack

Several scholars have suggested that the emerging postmodern society can be at least partly characterized by increasing social-psychological conditions of multiplicity and self-fluidity (Anderson 1997; Gergen 1991; Holstein and Gubrium 2000; Lifton 1993). Not surprisingly, numerous scholars of computer-mediated communication have emphasized and further explored this theme (see Benedikt 1994; Featherstone and Burrows 1995; S. Jones 1995, 1998; Porter 1997; Reid 1994; Turkle 1984, 1995; Waskul and Douglass 1997; Waskul, Douglass, and Edgley 2000). Yet, to suggest that the landscape of social and cultural relations has changed does not necessarily imply that collective ways of thinking about the social and cultural world change at the same time, rate, or direction. While change in one dimension of culture usually sparks changes in others, to be sure, social and cultural change is not monocausal or instantaneous, nor is it always totalizing. Even in conditions of major social change, older social orders and cultural constructs do not always readily dissolve, nor do they necessarily vanish altogether. Sometimes, older social and cultural orders tenaciously persist, even when experiential conditions no longer sustain, reinforce, or validate them. Indeed, "beliefs or traditions don't necessarily erode because they are flawed" (Gergen 1999: 29)—one need only consider the continued existence of the British monarchy to illustrate the point.

Experiential conditions of multiplicity and self-fluidity may have become increasingly common throughout society, yet cultural prerogatives for a unitary and singular self (i.e., "true self") remain dominant. That is, "we still search for a solid, predictable core of self, even though the conditions for the existence of such a self have long since vanished" (Hochschild 1983: 22). We may consider circumstances like these a *cultural crack*—a fracture in the structure of social reality defined by conditions in which cultural understandings of self and the world do not appropriate the experience of self and world. In short, a cultural crack is a misalignment between understandings of the world and the experience of it.

A cultural crack is similar to what William Ogburn (1964) calls a "cultural lag." To Ogburn, a "lag" refers to a condition where some cultural elements change more quickly than others, causing a disruption in the cultural system. These "lags," according to Ogburn, are precipitated by the invention, discovery, and diffusion of new technologies. In proposing the term "cultural crack," I do not necessarily disagree with Ogburn's concept, but I do take issue with its implied assumptions. Ogburn saw technology as the driving force of social change—he saw a world in which we are forever struggling to adapt to the cultural implications of new technology. That may often seem the case, but the influences of technology are not nearly as determinate as Ogburn's concept seems to imply. Even more, the term "cultural lag" suggests that one aspect of culture has progressed while another is lagging behind, further implying a kind of evolutionary directionality to Ogburn's conceptions of social change. By using the term "cultural crack," I intend to avoid this technological determinism and implied use of a progressive grand narrative. Given contemporary currents in social thinking, the metaphor of a crack seems more appropriate than a lag. A crack implies neither progress nor directionality—it is simply a disjuncture or misalignment.

Regardless, these cultural cracks appear to be an increasingly common feature of the experience of life in society. As an individual caught in the fault line of major social and cultural change, one need not look far to find conditions of growing uncertainty about the nature and meaning of life in society. "Everywhere there is questioning, challenging, mistrust, and resistance" (Gergen 1999: 30). Indeed, "living with ambiguity" (Bauman 1991) and contradiction has become a predominant feature of life in contemporary society. The relationship between these conditions and the experience of selfhood is apparent: "the purported tensions plaguing contemporary experience derive in part from the belief in a coherent self that is forced to confront society's myriad needs for identity" (Holstein and Gubrium 2000: 13).

Online chat environments often exemplify this struggle over meaning. Indicative of the conflict embedded in this cultural crack is the vibrant controversy that sometimes develops in response to these issues when they emerge in online chat conversations. Consider the following discussion that resulted from a single question:

Question	: Can you tell the difference between online fakes and those that are "genuine"?
Mary	: Yes I can, but it took over 7 years of being online
Ellen	: You are kidding yourself, Mary
Hank	: I'm a genuine fake
Bruno	: Sometimes I can spot the liars
Ellen	: How could you tell if I was lying? I mean, REALLY!
Siz	: The tone of your typing, Ellen. LOL!
Mary	: It's not hard to do I was a liar for years now I only tell the truth
Ellen	: Online is fantasy, if you want to know the truth, you have to meet someone. See their body language, look in their eyes.
Siz	: Right, online everyone is well endowed and good looking! LOL!
Mary	: Not to me. If people need to lie and only hurt themselves it's ok
Bruno	: See that's the problem. People making this a fantasy world

Bruno	: I don't have the need to lie online. It's classic projection—people project what they want onto a person, and disappointment awaits!
Siz	: I think alot of people use this as an outlet, saying or doing things they won't let themselves do in reality.
Ellen	: Yes, I think so too, and the behaviors become surprisingly extreme! I mean, every woman online says she loves oral sex. THAT has to be a lie!
Siz	: You are ALWAYS who you are, but some circumstances will censor or govern your behaviour differently.
Question	: Some people claim that when online the lack of inhibitions reveals their true self. Do you think that's true?
Bruno	: kinda like being drunk then?

Although a humorous discussion, it is clear that the participants in this conversation are expressing strong sentiments from widely divergent points of view. When these viewpoints collide, as this conversation illustrates, we are witness to a "reality-definitional contest" (Loseke 1987). Similar to patients in a mental institution (Goffman 1961) and shelter residents in a battered women's home (Loseke 1987), these chat participants are negotiating disparate definitions of themselves and experiences that are brought on, in this case, by the cultural crack between the prerogatives of a unitary self and experiences of multiplicity and fluidity. These reality-definitional contests are vibrant and emotionally charged because they refer to important systems of knowledge on which we base a significant portion of our self—they are, quite literally, the Goffmanian "pegs" on which we hang our self. Or, in less dramatic terms, these "definitions are powerful because claims-making gradually moves them into the general stock of knowledge" (Loseke 1987) that organizes and provides meaning to the experience of life in society.

These divergent views reveal the two primary interpretive means by which online chat participants negotiate the disparity between cultural prerogatives of a unitary self and experiences of multiple and fluid self-enactments. On one hand, there are those who embrace the postmodern aesthetic of multiplicity and self-fluidity. On the other hand, there are those who retain faith in independent, unique, and "true" selfhood but see their life in society as incompatible with the cultivation of such a self. Let's take a closer look at both.

Negotiating the Cultural Crack: Embracing a Postmodern Aesthetic

First, there are persons who wholeheartedly accept and embrace self-multiplicity and self-fluidity. Characterized by a rejection of the ideology of a unitary "true self," these people play with multiple potentials for being and see little difference between online and off-line social worlds. For example, consider the following:

Question   : Do you think most people are who they claim to be when online?
Response   : I think that they are showing one of their true selves. I think people
              have many selves. I don't necessarily think the people one meets in
              person are being any more genuine than those one meets online.

Each of my screen names is a different representation of my multi-faceted personality.... I consider myself witty, intelligent, and sophisticated. And I have a sparkling personality! Each screen name is just a representation of different parts of my personality. One screen name is the eternal child within me—she has the 64 pack of Crayola Crayons, you know, the one with the sharpener and stuff. Another screen name represents the lady I am—she has read Emily Post's Book of Etiquette, 15th edition. And a third screen name dreams of waltzing to Strauss on a moonlit eve in Vienna. *But I don't act any differently offline.* [emphasis added]

For these chat participants, it is apparent that the multiplicity of online chat environments presents no conflict with regard to cultural ideologies of selfhood. Clearly, these people have rejected the cultural prerogative of a unitary self and adopted an alternative view of the "true" nature and meaning of selfhood. This alternative negotiates the cultural crack by embracing the postmodern aesthetic of multiple and fluid forms of selfhood. Interestingly enough, the coherency of this alternative view is accomplished and maintained by what Berger and Luckmann (1966) call "conceptual nihilation."

In *The Social Construction of Reality,* Peter Berger and Thomas Luckmann (1966:104) identify and discuss various "conceptual machineries of universe maintenance." While these "machineries" are often thought of as institutional forces by which the legitimacy of the dominant social and cultural world is maintained, it seems reasonable to conclude that these same "machineries" may be employed to define and maintain the boundaries of alternative symbolic universes. Indeed, "all social phenomena are *constructions* produced historically through human activity" (Berger and Luckmann 1966: 106, emphasis in original), and therefore it is reasonable to conclude that they are all maintained by the same basic "machineries."

These participants "nihilate" (Berger and Luckmann 1966) the cultural crack by accounting for self-multiplicity in terms unique to their alternative views of selfhood. For example, as one participant blatantly states: "Truth does matter. You really can not be 100% sure." Another participant justifies this sentiment by saying: "What is the difference if I lie to every person I ever talk to?" Furthermore, conceptual "nihilation *denies* the reality of whatever phenomena or interpretations of phenomena that do not fit into that universe" (Berger and Luckmann 1966:114, emphasis in original). These participants neutralize the cultural prerogative of a unitary self by allocating it an "inferior ontological status, and thereby a not-to-be-taken-seriously cognitive status" (Berger and Luckmann 1966: 115). Therefore, as the previous quotes illustrate, the cultural prerogative of a unitary self is deemed foolish and mistaken. As one participant casually remarked, "my truthfulness borderlines on naiveness." Another participant questioned the ontology of truth in selfhood when she said: "I guess I try to be honest about myself. Why lie when you know everything may be false? *I have both, thank you*" (emphasis added).

These people embrace a postmodern aesthetic and embody what Lifton (1993) calls the "Protean self," but Gergen (1991, 1994) more accurately describes as a "relational self." Lifton argues that, due to contemporary socio-cultural conditions and historical forces, people have evolved into shape-shifting modes of selfhood that he refers to as the "Protean self" (after Proteus, the Greek sea god of many forms). However, Lifton views the Protean self as largely a tactical and selfish move that enables persons to engage in continuous exploration and personal experimentation. This is not what these chat participants indicate.

These chat participants display more than selfish and egoistic selfhood. Instead, these chat participants suggest a fully "relational self" (Gergen 1991, 1994). That is, these participants embrace multiplicity yet acknowledge that individual autonomy is a state of "immersed interdependence, in which it is relationship that constructs the self" (Gergen 1991: 147). These participants are quite comfortable, not with an orgy of egoistic selfhood, but with abandoning the view of self as an innate object (i.e. an innate, paramount "true self") and viewing these self-constructions pragmatically as a means of getting on in the social world. As aptly summarized by one participant:

Question   : How do you know if the people you meet online are genuine or fake?
Response   : Do we ever really know people, though we may meet them in person? Everyone is genuine in the moment. The moment is a spark. The person is genuine in many styles. He has many faces. Like a masquerade.

Negotiating the Cultural Crack: A Real Self in an Unreal World

In contrast, other chat participants take a different approach to negotiating this cultural crack. Whereas the first approach reveals an accommodation that nihilates the cultural prerogative of a unitary self (and therefore, poses no embedded conflict), this second approach embodies a worldview that equally captures the struggle for meaning that participants in chat environments may encounter—albeit in a strikingly different manner. These second negotiations consist of chat participants that accept that people *can* be self-multiple and fluid but deny that they in fact are. As one participant confidently states, "I am me, and that is all I am." Another participant explains: "I am who I am! I can't change that! I am always me." More explicitly, one participant refused to answer questions about online self-multiplicity on an e-mail survey. Seemingly offended by the questions, this participant bluntly replied with the following: "I see nothing special in your survey. I am me, and am what I am. If people don't like me for what I am, then it's their right. I don't put on a show for anyone. I am proud of who I am. Chatrooms are a place to go to make jokes—nothing more."

These participants tend to make sharp distinctions between online and off-line social worlds—it is clear to them that these two worlds are fundamentally different. By virtue of this distinction, these participants often perceive online activities as an escape from the realities of everyday life into an illusionary never-never land of online fantasy. In fact, for these participants, the cardinal attraction of online chat environments stems precisely from this perceived unreality. For them, online chat environments represent a place to take a socio emotional time-out—a conceptual vacation from the burdens of everyday life:

This is all make believe...a world of phony smoke screens, lights, and mirrors. But illusion is always prettier than reality.

Response	: Online is an escape from loneliness.
Question	: Why online, and not in person?
Response	: Real life is too scary in the real world. Too much commitment, sexually transmitted diseases, etc....We are all trying to escape from reality for a little while and this fantasy is a great way to do it.
Question	: Is that what it all amounts to—a fantasy?
Response	: It is an ESCAPE not fantasy.
Question	: Escape from what?
Response	: Work, bills, life.

As the above quotes suggest, these participants often apprehend their experiences of everyday life as confined within a social world that is stifling, alienating, overwhelming, restrictive, and inhibiting. The demands of everyday life are sometimes seen as burdensome chores to be endured. Locked within such an oppressive social world, even the task of face-to-face interaction can require too much of one's self. A condition of "multiphrenia" (Gergen 1991) comes to characterize their social lives, where the individual is split into a multitude of self-investments. Pulled in myriad directions, the maintenance of the modern self becomes an overwhelming burden:

> I find at times that it's just hard to listen, to have patience to discover someone, or get to know someone. I think it's cuz living just requires so much of you, that often you're just too tired to try.

> Sometimes you don't wanna talk on the phone. [Online] you don't have to hold the receiver, and you don't have to sound like you are paying attention. It's easier.

> I like to talk to people without anyone really knowing me.... It is nice to be able to have a conversation with a friend when you don't feel like physically talking. It's easier sometimes. It's like, you can talk to a lot of people at once, and say the same stuff you can on the phone.

Ironically, many of these people tend to view their online cyberself as more reflective of their "true self" than who they are off-line. That is, in spite of drawing sharp distinctions between online and off-line social worlds, and denying that they are multiple-fluid (i.e., "I am me!"), these chat participants often arrive at the interesting and seemingly conflicting conclusion that their online "cyberself" is more reflective of who they "really" are. Consider the following:

> Actually, in most chatrooms, if anyone is talking long enough their real personality comes out.... Once you get people talking *they become more real than face to face*— no masks (emphasis added)

> *I feel more like my true self online.* Is that weird? My whole life should be conducted online. I'm more confident. I'll say anything! You can say what you want without looking someone in the face. Sometimes it's easier to say what you think on here. Well—as my shrink confirmed—it's safe! You can open up to people (emphasis added)

> Have you ever thought that people are more open here, and can be with no inhibitions, and can actually get closer to knowing someone? I feel people can open up easier here. People are more open when they are not face-to-face

It is extraordinary that these participants can sustain such apparently conflicting beliefs. One chat participant explained, "The anonymity allows me to be myself, allowing a certain freedom of expression in which I rarely indulge in person." The irony of this statement is apparent: Online chat is anonymous (no one knows who you are), yet in that anonymity, this chat participant claims to be "more myself." Just how people can be "more themselves" in an environment where anyone can be anyone and no one knows who is who is unclear, if not paradoxical. What is interesting here is that, on one hand, these chat participants claim that they "are who they are—and no one can change that" when online; they are the same people as when they are off-line. On the other hand, they claim to be "more real" online than in face-to-face interaction. Most surprisingly, none of these participants sees these two statements as conflicting. Why? The answer rests in how these participants conceive of their self, social world, and the relationships between them—conceptions that they have already articulated.

It is clear that, unlike those who embrace the postmodern aesthetic, these chat participants see themselves as self-contained individuals. Who they are is something that is *within* them. These chat participants embrace and accept the cultural prerogative of a unitary self. From their perspective, inside of them they have a "real self." It is also clear that these chat participants see themselves as pulled in a multitude of incoherent directions that alienate them from their "true inner self." Thus, only by retreating from the burdens and demands of everyday life is their latent "true self" momentarily released from its social and cultural imprisonment. Those who embrace the postmodern aesthetic perceive the multiplicity and fluid forms of selfhood in online chat to be a *mirror* of the same kind of people we are in everyday life. These chat participants, however, perceive the multiplicity and fluidity of online chat as a *release* from the tyranny of their social and cultural alienation.

It is not difficult to understand this point of view so long as we think about it in the context of the perceived "rat cage" of off-line social life and the perceived "fantasy" or "unreality" of their online activities. In fact, most of us can probably empathize with how these chat participants conceive of the monotonous "rat cage" of the "real world," as vividly described by Stanley Cohen and Laurie Taylor (1992: 48, 49, 51):

> The regularized nature of our life begins to loom within consciousness as a cause for dissatisfaction ...Our house appears much like...[a] little box.... our relationship with our spouse and children indistinguishable from those paraded in the soap operas or radio and television. Our job comes straight out of the textbook discussions about alienation at work. We appear to live by order, moving from network to cable television, from vinyl records to compact discs, from natural gas to microwave, along the market tramlines of consumer society. How may we declare ourselves still free and indeterminate, individual and unique, when uniformity asserts itself so massively within our daily life?...At such times words like "freedom," "spontaneity" and "indeterminacy" seem empty slogans....The habitual stretches out like a contagion into every region of life; it feels inescapable. This is the world despaired of by the

existentialists, the empty hollow nothingness of a Beckett play, in which no one moves, nothing changes, and no one comes.

In these perceived conditions, which are similar to those described by the chat participants, Cohen and Taylor make clear how selfhood is stifled. One's very sense of self, and the materials by which persons construct a unique personal identity, can be seen as systematically eliminated by a perceived imprisonment within the monotony of the alienating experience of everyday life. As Cohen and Taylor (1992: 46-47) further explain:

> ...each day's journey marked by feelings of boredom, habit, routine. We feel dissatisfied with our marriage, our job and our children. The route we take to work, the clothes we wear, the food we eat, are visible reminders of an awful sense of monotony. For some people such feelings may be so intense that they are led to search for alternative realities; they set out to change their whole world. But for most of us, the periodic sense of dissatisfaction is related not to marriage, work, children as such; we do not wish to rid ourselves of these involvements altogether. What we object to is the sense that we are sinking into a patterned way of existence in all these areas; that they no longer appear to us as fresh and novel. They are becoming routinized. They no longer help us to constitute our identity.

When the social world is apprehended in this way, it is not unusual for people to take measures to reestablish a sense of individuality and unique selfhood. In fact, this is something that we already know. Through momentary "role distancing" (Goffman 1974), encompassing "adventures" (Simmel 1911[1971]), or dramatic "escape attempts" (Baumeister 1991; Cohen and Taylor 1992) we all, at times, create experiential spaces in which the routines and roles of a patterned everyday life may be temporarily abandoned so we may retain a modicum of individuality in the face of encroaching social structures. In fact, this is what Goffman (1961: 320) meant when he wrote:

> Without something to belong to, we have no stable self, and yet total commitment and attachment to any social unit implies a kind of selflessness. Our sense of being a person can come from being drawn into a wider social unit; our sense of selfhood can arise through the little ways in which we resist the pull. Our status is backed by the solid buildings of the world, while our sense of personal identity resides in the cracks.

For some chat participants, the experience of self on the Internet becomes precisely one of these means for detaching from a perceived stifling and oppressive reality. For these participants, online chat becomes an "activity enclave" (Cohen and Taylor 1992)—a liminal environment where they carve out space for the expression and experience of meaningful personal selfhood. Like hobbies, games, gambling, and sex (Cohen and Taylor 1992), online chat is an experiential space where participants dig out a safe place for meaningful personal activity and individualized selfhood; it is a momentary slip through the fabric of what is perceived as an oppressive and monotonous social reality.

Seen in this light, it is easy to understand how some chat participants can perceive their online cyberself as more reflective of "who they are" than off-line self-experiences. If one perceives the "real world" as stifling, monotonous, and predetermined, then the liminal world of online environments becomes "unreal" precisely because it lacks these qualities. Because of this perceived freedom, participants can enact and experience a form of selfhood that may appear more personally meaningful than that which is predetermined and imprisoned by the monotonous and burdensome routines of everyday life. Thus, these participants may or may not act differently online than off-line—that is beside the point. Their point of view may be ironic and contradictory—but that too is beside the point. Rather, they *perceive* the online world as allowing for more personalized self-enactments that are not (or cannot be) realized in everyday life, and therefore they perceive their cyberself as more reflective of their "true self."

Conclusions

All human societies are built upon a lie, the lie of self.... Having accepted the lie—which we all must, and all do—we live out our lives as the selves we believe ourselves to be.
—Walter Anderson, *The Future of the Self* (1997)

It is doubtful that online chat participants spend much time pondering the ontology of selfhood in relation to computer-mediated communication. It would be a mistake to conclude that anonymous online chat is teeming with such weighty cultural and intellectual discussion. Only among scholars are these kinds of conversations routine; only academians and moral entrepreneurs transform otherwise taken-for-granted experiences of life in society into problematic conditions of some perceived magnitude. This is not to deny that social life is filled with innumerable problematic conditions and contradictions—it surely is; after all, that's what makes it so interesting. The fact that most people do not normally ponder such questions only reveals the greatest source of society's power over the individual—the social world *is* taken-for-granted; we seldom think about it.

The negotiation of the cultural crack between prerogatives of a unitary self and experiences of self-multiplicity in online chat was precipitated by the very questions that I asked participants. I forced this disjuncture to the surface and made an issue out of what is normally a non-issue for online chat participants. This, however, does not diminish the significance of the cultural crack. Whether recognized or not, there is a misalignment between the cultural appropriation of the self and experiences of selfhood, and this disjuncture has important implications.

When faced with the challenge of negotiating this cultural crack in online chat environments, there emerges a conflicting set of dichotomized accommodations. It may be reasonable to suggest that these same accommodations are more common throughout the contemporary social and cultural landscape. Furthermore, it is reasonable to suggest that these interpretive approaches represent two poles in a continuum of possible negotiations. At one extreme there are those who fully embrace a "relational self" (Gergen 1991, 1994) and reject the cultural prerogative of a unitary self. At the other extreme there are those who retain faith in the ideology of a unitary paramount self, yet accommodate the disjuncture by viewing the world as incompatible with its cultivation. At the most general level, one form of negotiation rejects cultural definitions while the other rejects social conditions. In spite of their differences, both negotiations achieve the same objective of neutralizing the cultural crack by relocating selfhood within alternative symbolic systems that account for the disjuncture. In final analysis, neither is fully committed to the world in which we live. While those who embrace a fully "relational self" nihilate the cultural prerogative, those who retain faith in a unitary self do so by nihilating the social world (in this case, the social world is conceptually liquidated into an alienating tramline of monotonous, impersonal, and meaningless routines).

A single view of a cultural condition rarely fits perfectly and consistently with each individual experience and social situation (Williams 1976). Therefore, how individuals more generally negotiate the crack between cultural prerogatives of a unitary self and experiences of self-multiplicity is not likely to be simply characterized as either an embracing of the postmodern aesthetic or a salvaging of true selfhood. It would be another mistake to assume that the complex and multifaceted experiential conditions and understandings of people can be that easily dichotomized. In the lived experiences of life in society, it is likely that most people negotiate this cultural crack somewhere between these two ideal/typical extremes in varied ways in varied situations—more often than not, we combine elements of both to create personal configurations of the nature of our unique selves. In short, we artfully pick and choose from what is experientially and culturally available to craft articulations that make sense to our individual lives and experiences. Or, as Holstein and Gubrium (2000: 162, 179) state:

> The ways that culture is used—the fashion in which cultural categories are applied—is always variable and contingent...it yields variable images that reflect the artful combination of diversely available interpretive resources.... This combination of common resources, artfully articulated, gives us the possibility of individualized selves that nonetheless bear a striking resemblance in the way they are structured.

This last point is essential for understanding the negotiations of the cultural crack that are implicit in online chat environments, and the significance of them for better understanding the lived experience of life in society. Increasingly, the normative experience of life in society is not that much different than online chat environments. Like everyday life in contemporary society, the online chat environment provides tremendous opportunities for relatedness, but in a way that also offers tremendous opportunities for individuality. The recognition that our selves exist only within relationships among others undermines the ideology of the independent unitary self, yet increased individuality is a necessary and sufficient condition for retaining faith in the uniqueness of one's self. Opportunities for self-presentation are greatly enhanced while opportunities for "altercasting" (Weinstein and Deutschberger 1963) are greatly diminished. In short, the ideology of individualism remains the cultural bedrock to our understandings of personhood, within the same social conditions that would seem to deny the development of singular, stable, unique, and independent selves. Online chat illustrates this well, and it is reasonable to suggest that the same overall conditions increasingly characterize our lives *off*-line.

NOTES

1. By "situationally variable," I intend to concede to the often cited fact that online chat between "real-life" friends, work associates, or even among those who interact online frequently over a long period of time is not likely to be characterized by the same generalizations described for anonymous online leisure chat. My interests are in anonymous online leisure chat. Furthermore, these kinds of distinctions are essential for an adequate understanding of computer-mediated communications. Acknowledging the interrelationship between *purposes* and *conditions* of communication and communicative outcomes is essential to avoiding the trappings of technological determinism—an oversight that characterizes early studies in computer-mediated communication (see Baron 1984; Kiesler et. al. 1984; Rice 1984, 1989; Sproull and Kiesler 1991; Walther and Burgoon 1992). Ultimately, as I have and will continue to say repeatedly, the nature of computer-mediated

communication is like all other forms of human interaction—it is *emergent* from the purposes of the communication, within *situated* contexts (in this case, a computer mediated-environment), that are *negotiated* with others.

❖ CHAPTER FOUR

Text Cybersex:
Outercourse and the Subject Body[*]

> The body—or its absence—is central to contemporary notions of "cyberspace," "the Internet," "virtuality": computer-mediated communications (CMC) are defined around the absence of physical presence, the fact that we can be interactively present to each other as unanchored textual bodies without being proximate or visible as definite physical objects.
>
> —Don Slater, *Body and Society* (1998)

In the 1990s, due primarily to its novelty and marketing potential, "cybersex" has rapidly become a catchall term used to refer to an enormous range of computer-mediated sexually explicit material. It is not unusual for the word to be associated with adult CD-ROMs, sexually explicit interactive games, and pornographic sound, image, and video files available on the Internet (Robinson and Tamosaitis 1993). Indeed, if one searches for "cybersex" on the World Wide Web one can't help but be impressed (for better or worse) with the plethora of smut peddlers eager to sell access to a hearty collection of kink. However, when we strip away the marketing exploits, it turns out that cybersex is only marginally related to pornography, and is more squarely situated in a specific kind of interactive erotic *experience*.[1]

Among those familiar with this sort of activity, there is little ambiguity as to what constitutes cybersex. Cybersex strictly refers to erotic forms of real-time computer-mediated communication. Through typed text, live video, and sometimes spoken voice, cybersex participants meet one another for erotic encounters in the ether of computer-mediated environments. Regardless of the form, however, cybersex always entails meeting someone else in these decidedly liminal "places" where real people communicate with other real people for the purposes of explicit sexual communication, arousal, and/or gratification.

---

[*]An early version of the contents of this chapter was coauthored by Mark Douglass and Charles Edgley, published in *Symbolic Interaction* (2000, Volume 23, Number 4, pages 375–397).

More than just interactive and communicative, cybersex is unique in how it blends an experience that is normally quite embodied in a form of interaction that is distinctly disembodied. This contradiction may be the most extraordinary characteristic of cybersex. For most people most of the time, sexual intercourse represents an ultimate in embodiment—a necessarily corporeal experience in which physical bodies interact. Indeed, in order to have sex, even if only by one's self, it is necessary for there to be a body for one to have sex with (at the very least, one's own). The consequences of these corporeal sexual encounters evidence themselves in bodily matters— sexual intercourse is wet, odoriferous, parts of the body swell and engorge— the whole experience teems with various organic fluids. For most people sex is all about these corporeal bodies, organs, and fluids; even the president of the United States once learned that the presence of these fluids—on a dress or anywhere else—is primafacie evidence that bodies once had sexual relations. By contrast, "netsex is an incorporeal medium and the essential ingredient in sexual activity, the body, is left out" (Leiblum 1997: 25). Thus, cybersex is a distinctive kind of sexual activity, one that plainly contradicts the conventional definition of sex.

Outercourse: The Body in Sex and Cybersex

In any form of interaction, the body is an object to one's self and others. That is, the body is always both a physical, tangible, corporeal object that may be seen and acted upon by one's self and others, while it is also a subject that is experienced and filled with various social and cultural meanings. It is in how we uniquely manipulate and configure relationships between the appearance of our body as an object and the experience of our body as a subject that we achieve a distinct embodied self. In short, through an extensive process of discursive and non-discursive communication, the body is always presented, decorated, manipulated, and negotiated as an object as one accomplishes and experiences one's embodied self in situated environments of social interaction.

In sex, this dual character of the body is unambiguous, making it not only a distinct embodied experience but also among the most pleasingly embodied. After all, lovers have sex with each other's *bodies* that just so happen to have selves attached to them, and, conversely, lovers have sex with each other's selves that just so happen to have *bodies* attached to them. "Even in a long-term romance there is a world of difference between the desire for the *lover's* body and the desire for the lover's *body*" (Tisdale 1994: 10, emphasis in original). As Georg Simmel (1950:131) suggests, "sexual intercourse is the most intimate and personal process, but on the other hand, it is absolutely general...the psychological secret of this act lies in its double character of being both wholly personal and wholly impersonal." For these reasons, sex is an ideal context for examining the relationships between the objective and subjective qualities of both bodies and selves and how these are mediated by situated social interaction in a process of communication.

Cybersex is a form of experience that makes these relationships especially salient. After all, the Internet is among the most dislocated and disembodied contexts for real-time human interaction. On the Internet, participants interact with others in electronic space that is quite separate and removed from the immediate presence of the body. Because sex requires a body, cybersex participants must therefore *evoke* one, and in this process, the role and nature of body is not only magnified but also transformed.

By pressing the embodied experience of sexual intercourse through the disembodied medium of computer network technologies, a McLuhanesque transformation alters it into something different—an experience we may call sexual *outercourse*. In outercourse, images and/or words fully replace the corporeal body as they are crafted among participants to represent the whole of sexual and erotic interactions between them. In cybersex, participants do not interact directly with the bodies of their cyber lovers. Instead, all they see and interact with are the words and images that *represent* them. While the pleasures of sexual intercourse are encapsulated in physical contact between corporeal bodies, the pleasures of outercourse are encapsulated in dislocated and disembodied erotic communication where participants latently rearrange taken-for-granted relationships between bodies, selves, and situated social interactions.

Because sex is a necessarily embodied experience, it presents an ideal lens for examining relationships between the body, self, and situated social interaction. Because of the disembodied nature of the Internet, these relationships are especially evident. For these reasons, cybersex is a precariously situated experience that poses a strategic context for examining relationships between bodies, selves, society, and how people manipulate these relationships within this unique form of interaction to create new configurations of erotic experience. The pressing questions here, then, are: What are the relationships among bodies, selves, and society within these forms of erotic computer-mediated social interaction, and what can these relationships tell us about bodies, selves, and society in everyday life?

In this chapter, we will examine these questions in the context of text cybersex. In text-based cybersex, typed words come to represent and replace interactions between corporeal bodies. Like phone sex, text cybersex is purely communicative: there are no co-present bodies, actions, touches, and—unlike phone sex—there are no spoken words; the whole of the experience is communicated in detailed and intimate text that represents appearances, actions, touches, and utterances. Thus, whereas normally the body is both a viewed object and experienced subject, the semiotic bodies evoked by participants in text cybersex are only subject bodies. The corporeal object-body is unseen, dislocated from an emergent subject body that is interactively crafted by participants in a process of communication. What is the nature of these semiotic subject-bodies of text cybersex and what can they tell us about the relationships between selfhood and bodies in everyday life?

Self, Body, Society, Sex, and Cybersex

> In everyday life and commonsense thought, our bodies and our selves are more or less the same thing.
>   —Walter Anderson, *The Future of the Self* (1997)

Few will deny that sexual intercourse is a form of social interaction that is necessarily related to the body. Yet even so, it is also widely recognized that sexual interaction between bodies has neither a fixed nor necessarily normative state. Sexuality can assume a stunning range of expression between individuals, making it an extremely complex and multifaceted human experience. Part of what makes it so is the fact that sexual expression is located at the intersection of a triadic relationship: the selves that we are, in relation to physical bodies, situated in a socio cultural context, represent three interrelated elements that together form the core of any sexual encounter. These three interrelated elements converge to form body-to-self-to-society relationships that are explicit when considering human sexual activity. Yet, each of the components of this body-self-society relationship is fundamentally different from the other.

The self is a symbolic referent of an individual—a fluid system of meanings that refers to the person. These meanings emerge only through communication and interaction with others in the context of particular social situations, roles, and encounters. In other words, a self emerges only as one interacts and communicates with others and the meaning(s) of those actions are transferred to the person. Selfhood is multiple and dynamic. A self can change (or be changed) as one moves from situation to situation, role to role, place to place, developmental stage to developmental stage. All persons enact a wide range of selves, as we are one thing to one person and something else to another.

The body, on the other hand, is not simply a fluid set of meanings. Unlike selfhood, the body clearly manifests itself in empirical qualities that occupy space and time. Consequently, the physically verifiable corporeal body has at least two important functions for traditional conceptions of selfhood. First, because selfhood is multiple, dynamic, and fluid, "the nebulousness of personal identity has caused it to be commonly conceived in concrete form as coextensive with the physical body" (Davis 1983: 112). In other words, it is to corporeal bodies that we commonly associate or affix systems of meaning that collectively comprise the self. Or, as Goffman (1959:253) describes, the "body merely provide[s] the peg on which something of collaborative manufacture will be hung for a time." Without a body, there is no "peg" on which to associate or affix a stable set of meanings that we may refer to as a comprehensible person.

Second, the corporeal body is not only the stable "thing" to which we associate fluid systems of meaning that comprise the self, it is also commonly seen as that which contains the self. In spite of the fluidity of selfhood, the fixed and verifiable existence of a physical body has always posed limits on the range of multiple selves that an individual may enact. Although the human body may be decorated and otherwise altered along a seemingly infinite continuum, the physical body remains an important limiting component of self-to-social-world relationships. Normally, the physical body makes it difficult (sometimes impossible) for a man to present himself as a woman, a Caucasian as an African, a boy as a man, a woman as a girl, and so on. Although bodies may be modified and self-enactments may vary greatly, the physical existence of one's body poses certain limits and socio cultural restraints to the range of individual multiple selves.

In short, the relationship between self and body is most often seen, as Alan Watts (1966:ix) describes, as a "separate ego wrapped in a bag of skin." "Your skin serving as the living frontier where your self leaves off and your environment begins" (Anderson 1997: 67). Thus, the body is there as the most basic and essential element of selfhood—the stable bedrock to one's sense of being, who we think we are, and what others attribute to us. "Being a *body* constitutes the principal behind our separateness from one another and behind our personal presence. Our bodily existence stands at the forefront of personal identity and individuality" (Heim 1991: 74, emphasis in original).

While there is a clear relationship between selfhood and one's body, there is also an important relationship between one's body and society. As suggested by Allucquere Stone (1995), the body is the unambiguous core of taken-for-granted conceptions of a comprehensible person and politically recognized citizen. For this reason, self-to-body relationships are best understood and interpreted within the context of broader body-to-society relationships that exert powerful and independent influences of their own. For example, a loose interpretation of Michel Foucault's *Discipline and Punish* (1979) and *History of Sexuality* (1978) suggests that body-to-society relationships are significantly altered in conjunction with the emergence of new interpretive discourses within shifting power relationships in society. *Discipline and Punish* (1979) examines a shift from punishments that took the form of abuses to the body (i.e., hanging, decapitation, burning, etc.) to those restrictions of freedom and liberty that inflict suffering on the self. *Discipline and Punish* illustrates the emergence of a new discourse for body-to-society relationships that induced a change from body-oriented perceptions (i.e., the body as the whole of the person) to mind-orientations (the mind as the whole of the self). Similarly, Foucault's *History of Sexuality* (1978) illustrates how the production of an ever greater quantity of discourse about sex resulted in important shifts in the intersections and influences between self, body, soul, and society.

This kind of general understanding suggests that selfhood is negotiated in Goffmanian "cracks" (1968) that are, in part, pinched between the different relationships of one's body to one's social world. Experiences of selfhood emerge, are negotiated, and become validated as one's body enters the scene of interaction in the context of pre-established socio cultural systems of meaning. This process occurs in the framework of a triadic body-to-self-to-society relationship. That is, selfhood is contained by and affixed to the corporeal body, which is acted upon by society and interpreted according to prevailing societal discourses. Indeed, bodies represent a fundamental element of personhood, the experience of which is caught in the precarious margins between body-to-society relationships.

In short, the body represents the grounded referent of selfhood. The body is a necessary condition for all of action and interaction (Strauss 1993). "It is the medium through which each person takes in and gives out knowledge about the world, object, self, others, and even about his or her own body" (Merleau-Punty 1962[quoted in Strauss 1993:108–109]). The body is a medium of action and interaction for the self, an agent to the self, the object of action, and the fulcrum of societal reaction.[2]

However, as we have already learned, in the social worlds of online computer-mediated environments, there are no corporeal bodies. In spite of millions of participants and a plethora of pornography on the Internet, there are no bodies—only symbolic representations of bodies (at best). In online communication environments, a corporeal body is usually necessary to access and interact with others. However, in these communication environments—especially online chat—the body is necessarily left at the keyboard, behind the scene of the interactions that transpire. In most online communication environments, participants are literally disembodied. Because no physically verifiable or empirically measurable bodies exist in these communication environments, actions and interactions are entirely disassociated from the corporeal body. Thus, in these online social environments, bodies are transformed into symbols alone—representations, images, descriptive codes, and words of expectations, appearance, and action. In these environments, the activities of participants and experiences of self are neither contained by nor affixed to corporeal bodies. Both bodies and selves exist only as socially constructed representations—sets of meanings that emerge in a process of interaction.

Because the emergence and development of the self can be historically followed through changes in the means by which it is produced (see Gergen 1991; Stone 1995), this kind of transcendence of the corporeal body is a techno-social development that is especially well suited to the examination of contemporary experiences of personhood. Furthermore, because sex is among the most embodied of all imaginable activities—an activity that revolves around an interplay between bodies—sex is the peg on which I hang my brief examination of the body-self-society relationships that are made manifest in cybersex. This chapter examines the very means by which experiences of "self" and "body" are produced and operate in online social environments of a sexual nature. In short, this chapter is cast in the context of cybersex but is about the negotiation of processes between individuals, communications with others on the Internet, and the physical bodies that may or may not be grounded in an emerging matrix of virtual experience.

A Dramaturgical Approach to the Problem of Virtual Reality

Many contemporary scholars have explored troublesome questions in assessing what is "real" with regard to the unique situations posed by electronic media environments. Scholars of computer-mediated communication have noted how these problems are magnified in Internet communication environments, giving them distinctive qualities that we may call "virtual."

Although it has become common practice to refer to computer-mediated environments as virtual, this word often mystifies these environments and overlooks the complex nuances of virtual reality. Because "virtual" sex lies at the heart of the contents of this chapter, it is necessary that we clarify its meaning.

For our purposes, Brenda Laurel (1993: 8) offers a productive stance on the problem of framing virtual experiences:

> The adjective *virtual* describes things—worlds, phenomena, etc.—that look and feel like reality but lack the traditional physical substance. A virtual object, for instance, may be one that has no real-world equivalent, but the persuasiveness of its representation allows us to respond to it *as if* it were real.

To Laurel, virtual things are persuasive representations that contain qualities that allow persons to respond to them as if they were real. "Virtual" merely refers to things, situations, and experiences that are dislocated from the frame of the empirically real—they do not necessarily draw referent from nor are they necessarily a part of that which can be empirically verified or subject to direct measurement. From this perspective, the reality of virtual things is emergent from interactions with the representation, not a quality of the things themselves. This approach to virtuality is quite similar to dramaturgical social reality.

We may borrow from Goffman (1959) to suggest that the reality of online environments is a product of a scene that comes off, not the cause of it or a quality of the scene itself. Or, we may borrow from W. I. Thomas (1966) to suggest that the things of virtual environments are persuasive representations that become real in their consequences. Like elements of social reality, the "things" of virtual reality may not have an objective or empirical manifestation, yet they pose a persuasive representation that exerts real influence and allows people to respond to them as if they were real. When responded to as if they were real, virtual things assume a pragmatic and experiential reality that transcends the frame of the empirically real. Indeed, whether virtual or otherwise, all realities are far more than a mere collection of things—reality is something that is produced and experienced; it is this experiential process of production that is important.

Approaching cybersex within this framework, this chapter is not directly concerned with empirically "real" persons, "real" experiences of sexual arousal, or the orgasms behind these virtual sexual trysts.[3] Rather, this chapter seeks to understand how persons create sexual encounters that have the privilege of being responded to as if they were real and the bodies and selves that emerge in these circumstances. What is important is that online environments dislocate the physical body from the context of social interaction—in this respect, they are "virtual" encounters. By removing the frame of the empirically real, online communication environments allow for the enactment of new forms of selfhood and provide insights on new relationships between bodies, selves, and social situations. The key question is: What are the relationships among bodies, selves, and social interactions when selfhood is situationally freed from the empirical shell of the body?

Cybersex: The Simmelian Adventure of Outercourse

> Perhaps not since the Middle Ages has the fantasy of leaving the body behind been so widely dispersed through the population, and never has it been so strongly linked with existing technologies.
>
> —Katherine Hayles, *October* (1993)

Cybersex is a form of co-authored interactive erotica (Reid 1994). Text-cybersex is a written conversation by which participants transform a computer-mediated communication environment into a personalized interactive arena for sexual experience. In the liminal and anonymous space of online chat, people engage at least one other participant and type erotic actions, utterances, feelings, and happenings to one another. Like phone sex, cybersex is a process of provoking, constructing, and playing out sexual encounters through a single interactive mode of communication (see Flowers 1998). Also like phone sex, participants draw from a vast repertoire of socio-cultural symbols to construct a drama that compresses large amounts of information into the very small experiential space of a text medium (see Stone 1994). The enormous range of bodily sensations that typically accompany sex (gestures, appearances, expressions, odors, utterances, and physical sensations) are compressed in and exchanged through a single channel of densely communicated interactive text. Consequently, as many participants indicated, "good" cybersex requires a great deal of sexual and communicative literacy:

> An active imagination and expansive vocabulary help. Using predictable expressions is a little ho-hum. Just saying "I want to suck your dick" is unlikely to arouse many people.

> There are only so many ooohs and mmm hmmms you can type.

Cybersex occurs in online chat environments that are widely acknowledged as relatively anonymous forms of interaction. Participants choose levels of anonymity and are selective in the personal information they report to others. Within this generally anonymous form of online interaction, cybersex participants feel little need to anchor themselves to a physically fixed manifestation of self. Not surprisingly, numerous participants strongly suggest that the anonymity of online environments allows for a fluidity of self-presentations that is an important element of cybersexual experiences:

> Cybersex allows the freedom of sexual expression. Cybersex allows a person to be whoever or whatever they want to be!!

> It's erotic, it turns me on—the mystery of it. Not knowing who is really on the other end is really erotic—you can be anything. I may stretch truth, and live out fantasies...it allows you to be with whoever you want—no inhibitions.

Similar to online chat and other forms of role-playing and vicarious experience, the anonymity of the Internet allows participants to play at various selves in the drama of a socially constructed virtual situation. In this case, participants may assume a wide variety of roles in the enactment of an interactive sexual drama. This is an important element of the eroticism of cybersex—it allows participants to playfully toy with alternative vicarious experiences:

> Sometimes I pretend I'm a woman, I've also invented experiences (like 3 somes). ...Cybersex enables me to play out fantasies.... It allows you to take your dreams one-step closer to reality.

> You can do anything you want and you can picture anybody you wish.

> It's not real. People can take any identity they want, and they do. People lie about who they are to create sexual illusions.

As an added bonus, if the adventure is deemed unpleasant or uncomfortable, the screen name can be deleted and a new one created. This allows a participant the power to be virtually reborn under another alias, to start completely anew, divorcing one's self entirely from the old screen name and what had been done with it:

> I deleted my previous screen name because I tried something that went beyond my normal comfort range.

It's more anonymous. And you can disappear much easier if it doesn't work well.

Anonymity in conjunction with the power to delete and recreate screen names contributes to the emergence of a social environment where participants feel free to experiment with new social roles and presentations of self. Participants tend to perceive these interactions as a kind of "self-game" where anonymous leisure interactions become a form of recreation and communication play. As one participant stated, "We enjoy it more than some folks enjoy bridge! So, what's the big deal? Its merely another form of entertainment."

In cybersex, like online chat, participants interact with others from a wide array of socially constructed personae, with no necessary commitment to that which is verifiable. This observation, however, is nothing new. Not only have numerous scholars examined the fluidity of selfhood in the context of cyberspace (see S. Jones 1995, 1998; Meyrs 1987; Reid 1991, 1994; Stone 1995; Turkle 1984, 1995), but these kinds of fluidity and self-multiplicity are not unique to the Internet. A "cyberself," like any self, is situationally defined. Persons have as many selves as they have meaningful situations within which to interact (Mead 1934). What makes online environments unique is how the fluidity of self-enactments is expanded exponentially by the absence of a physical body.

When online, one is not only free to enact a multiplicity of selves but also able to enact selves that are beyond an individual's range of possibility due to constraints normally imposed by the physical body. As a result, cybersex and online chat present opportunities for a hyperfluidity of self-enactments due to the ability to transcend gender, skin pigment, age, weight, and all other socially meaningful characteristics of the physical body. In this context, all "fixed" bodily features become self-selected variables—potential components in a selective enactment of a self, not taken-for-granted constants or givens.

Yet, to say that cyberselfhood has the potential for hyperfluidity does not lessen its experiential importance to participants. A self is, by its very nature, symbolic—a necessarily fluid and situationally defined system of meaning.[4] Thus, there is no reason to conclude that cyberselfhood is any less meaningful than other forms of self-enactment—perhaps more transitory, ephemeral, or liminal—but not necessarily any less meaningful. In fact, through co-authored erotic fantasy with anonymous others, many cybersex participants claim to learn new sexual techniques and turn-ons, and vicariously experience arousal in ways that they would not (or could not) experience in "real" face-to-face sexual encounters. Consequently, the experience is not only meaningful but also sometimes highly valued:

> With cybersex I learned about stuff I didn't know, like maybe how to do some things better. Everyone should try it!

> Since I've started chatting with people online, I've been walking around in this perpetual state of arousal! It's wonderful! I mean, perpetual, never ending, I'm always thinking about sex, coming up with new ideas, listening to other peoples' fantasies and expressions and learning things I never knew existed!

Although cybersex can be described using terms such as "virtual" or "fantasy," and may be considered experientially liminal, often participants insist that the experience is meaningful and highly valued. In other words, "to say that virtual sex involves highly self-conscious role-play does not mean that the roles are in any way necessarily false" (McRae 1996: 83). On the surface, this appears paradoxical. However, this apparent paradox makes perfect sense when one considers the degree to which cybersex represents a Simmelian "adventure," and how both sex and fantasy are important enclaves in which persons create "free areas" for self-expression and identity work.

Georg Simmel (1911 [1971]: 189) describes "the erotic" as a supreme example of "the adventure" as a unique form of human experience. To Simmel, the adventure is a form of experiencing that involves "dropping out of the continuity of life...in contrast to that interlocking of life-links." Although an adventure falls outside the context of life, it remains connected to it. "It is like an island of life which determines its beginning and end according to its own formative powers and not—like the part of a continent—also according to those adjacent territories." To Simmel, "the adventure" stands outside the context of taken-for-granted everyday life, yet by being outside of everyday life, this form of experience allows a person to synthesize, comprehend, and provide new meaning to the everyday world that "the adventure" stands over and against.

Stanley Cohen and Laurie Taylor (1992) clarify and extend upon Simmel's theory by illustrating the ways that sex and fantasy culminate in "activity enclaves" by which people cultivate a safe place for identity work apart from the routinized realities of everyday life. Sex and fantasy can be "cultivated as a free area when it is regarded as a portion of life in which we feel ourselves free of the routine nature of the rest of existence...regarded as an activity in which we may 'be ourselves' or 'get away from everyday life'"

(Cohen and Taylor 1992: 125–126). Cybersex certainly can constitute one such activity enclave. The anonymous nature of online leisure interaction, in conjunction with both the fantastical and sexual dimensions of cybersex encounters, creates situational conditions that mirror Simmel's form of adventure and thus potentially create free spaces for the kind of identity work that Cohen and Taylor describe. Hence, cybersexual experiences may become quite meaningful and highly valued by participants, not in spite of their virtual and fantasy elements, but because of them. One participant states very clearly how cybersex provides an experience of Simmelian adventure—an "activity enclave"—in which virtual experiences provide a context for renegotiating self:

> Whether a guy or a girl sends me a private message and wants to talk, it's usually very exciting. I am 32 years old and think I am only now reaching my sexual prime, and I don't know that I'd have discovered certain things about myself without it. I never thought I could be so free with my emotions and fantasies, and it's even spilled over into my real life, I mean, now I feel free about talking about my sexuality (bi-sexuality) with other people openly, now that I've discussed it with myself first (which basically is what I'm doing here, talking to a nameless, faceless person, i.e., ME!).

Most often, cybersex resembles a Simmelian "island of experience" that is related to everyday life through participants' appeals to physical bodies—as a novel masturbatory innovation, a way to learn about sexual techniques, or a means for examining one's sexuality and the sexuality of others. Yet, in spite of interpreting cybersex through appeals to the physical body, it is the absence of physical bodies that promotes the perception of cybersex as a safe form of communication play, uncomplicated by the emotional baggage of face-to-face sex, enjoyable, and full of perceived therapeutic overtures:

> For me, cybersex is an opportunity to give someone else stimulation and fulfillment in about the safest way there is right now—no commitments, no diseases, just good clean nasty fun.

> Acting out of fantasies can be very healthy and therapeutic.

> It can actually help your real sex life. It helps you do things that you might find difficult to do in real sex. Try things out. Actually find out what the opposite sex likes.

> I guess the reason I do is because it is a safe medium by which to explore sexually. To experiment with those aspects of sex that you have not yet explored. To enhance

your sex life through the use of new ideas that are learned with a new sexual partner, without risk. It is also a way to be excited sexually without the performance anxiety that is present in face-to-face encounters. It is a way to express yourself sexually in a way you may not feel comfortable doing in a relationship.

The virtuality of the context and hyperfluidity of self-enactments allow cybersex participants to engage others in sexual self-games that are (by intent) related to and interpreted in terms of the physical body (i.e., the sexual arousal of them). Thus, the basic triadic body-to-self-to-social-world relationship is upheld in the context of cyberspace. In other words, in spite of the lack of physical bodies, cybersex is still a body-game enacted by participants according to prevailing socio cultural interpretive discourses. As one respondent stated, "It's a paradox. People say that what they like about [cybersex] is that people are not judging them by their appearance, but after age/sex checks, it is the first thing everyone wants to know." However, this begs the question: If cyberselves emerge situationally apart from and not necessarily related to physical bodies, to what is this self affixed in the absence of a corporeal body?

Text Cybersex and the Social Production of the Virtual Body

...differences among people are often most visible when the differences are least present, and most present when they are least visible.
—Joshua Meyrowitz, *No Sense of Place* (1985)

The hyperfluidity of selfhood, made possible by the absence of a corporeal body on the Internet, does not eliminate the important role of the body. In fact, the body remains quite important. By far the most common phrase in online chat is some version of the question "Are you a male or female?" The second most common occurring phrase is usually one of the following: "What do you look like?" "How old R U?" or "Where do you live?" Ironically, by means of this complex and technologically sophisticated network of communication, people are asking some of the most basic and fundamental body questions imaginable—questions that through embodied face-to-face interaction are not only unnecessary to ask, but typically quite inappropriate. One would be hard pressed to find another context where it is common and acceptable to ask people if they are male or female, what they look like, and so on.

A more important difference, however, is that in text cybersex, the body is entirely a subject. Since the object-body is nowhere to be seen, how a person "looks" to another online participant is not determined by physical presence. Rather, appearances are entirely dependent on information participants choose to disclose. "The majority of one's correspondents in cyberspace, after all, have no bodies, no faces, no histories beyond what they may choose to reveal" (Porter 1996: xi). Or, as one participant said, "what you read is what you get," which is an interesting contrast to bodies in everyday life where, normally, what you *see* is what you get. In this way, because the body is freed from any necessary or verifiable physical manifestation, it is transformed into complete symbol—a virtual body that exists only in words that are emergent in communication and detached from the frame of the empirically verifiable. Both bodies and selves become systems of meaning that surface in a process of communication and associated with whatever semiotic performance participants are currently enacting. The corporeal object-body remains at the keyboard—behind the dramaturgical scene. When the corporeal body disappears in this way; it is no longer an agent to which selfhood is affixed. In short, the body is no longer directly connected to the enacted self.

These conditions often lead people to believe that they are more free. Once released from the socio cultural shackles of the physical body, people often assume that cultural and social meanings associated with bodies somehow magically vanish (i.e., presumptions about gender, race, obesity, ugliness, etc.). Or, in other words, it is often believed that the basic triadic body-to-self-to-society relationship is reduced to a dyadic body-to-self relationship—with the self in complete control of (and unlimited by) the appearance and actions of the body. This kind of release from the tyranny of selfhood is a common theme in the rhetoric surrounding the medium. A 1996 MCI television advertisement promoted this seductive vision: "There is no race. There is no gender. There is no age. There are no infirmities. There are only minds. Is this utopia? No, the Internet."

This is a happy vision of egalitarianism that probably sold a lot of Internet service and remains in common currency in the rhetorical harangue that often surrounds major technological innovations. As Joshua Meyrowitz (1985:317) once observed, "some people mistake any discussion of the 'breaking down of boundaries among people' for a prophecy of a utopian society of harmony and bliss." Thus, we should not be surprised that "much of what has been claimed for cyberculture *is* overtly utopian" (Featherstone and Burrows 1995: 8, emphasis in original). However, we should also not be surprised that this egalitarian vision does not stand up to empirical muster, nor does it make reasonable sense. It is more reasonable to suggest that because online chat participants can present any body they choose, they will be more likely to do so in a manner that supports the social and cultural mandates of the situational self they are currently enacting. These online self-enactments are dramaturgical performances that do not materialize out of thin air. Rather, they occur on a metaphorical stage that produces scripted socio cultural performances to which participants tend to adhere. Therefore, we should expect the performances of virtual bodies to emerge as part of participants' presentations of self, and in highly consistent forms. In short, as communicated elements of a self-enactment, bodies are more likely to adhere to cultural and social prescriptions appropriate to the situation—not as a variable, but as a constant prerequisite. In this sense, bodies are not more free in the disembodied online context, but clearly more confined.

In text cybersex, regardless of how many words are exchanged, participants never so much as see each other. An in many conventional and ordinary experiences of everyday life (e.g., religion, love, solidarity, etc.), in text cybersex, self, body, and the whole scene of interaction emerge from a shared consensual hallucination (Gibson 1984) that is substantiated and validated within dramaturgical performances. In text cybersex, each participant contributes to the performance of the other within a negotiated agreement of what is desired, expected, and/or required of the situation. Or, as one respondent succinctly stated, "looks and communication all tie together." Thus, the disembodied context enables participants to sidestep cultural specifications of beauty, glamour, and sexiness, but it does not subvert these concepts (Reid 1994). The fluidity of both body and self-presentation does not free participants from the shackles of the beauty myth, but only allows persons to redefine themselves in accordance with that myth. As Elizabeth Reid (1994: 64) explains:

> When everyone can be beautiful, there can be no hierarchy of beauty. This freedom, however, is not necessarily one that undermines the power of such conventions. Indeed, such freedom to be beautiful tends to support these conventions by making beauty not unimportant but a pre-requisite...free from the stigma of ugliness not because appearance ceases to matter but because no one need be seen to be ugly.

Because participants can present a virtual body that supports a cyberself enactment, and because these enactments are grounded in culturally prescribed standards of beauty and sexiness, it should not be surprising to observe a conspicuous absence of fat, ugly persons with pimples, small breasts, or tiny penises. Consider, for example, these typical descriptions of self and body that participants on a commercial online system anonymously reported:

> I have brown hair, blue eyes, average height, average build, bigger-than-average cock!

> I'm 22, 6'0" tall, about 176 pounds, long brown hair (mid back), Good shape, and love to have a good time. I'm not stuck up, but I am very attractive

> My hobbies include workin' out; I have a 46"chest, 32"waist, and 22" biceps/great ass nice and firm and a thick 9" cock

The above represent typical and believable body descriptions. However, there is no reason why participants have to claim either actual or typical appearances. Persons can present a virtual body that is strikingly attractive, with hyperbolic sexual organs and absolute prowess in sexual techniques. Take for example these amusing descriptions:

> My hobbies include using my 13" LONG 4" THICK Penis on Women. Selectively meeting attractive Women and sexing them with my 13" penis

> I'm a 21 year old single female 5'7" with blue-gray eyes. 124 lbs, 44DD-28-30.

> I'm 5'7, Long Black Hair, Brown Eyes, 46DD-30-36, 125 lbs.

> I am 5'2, 110, blnde/brn waist length hair, green eyes, 48DD

Granted, it might be possible, as one of the above participants suggests, to have a thirteen inch penis that is four inches thick. However, such a penis is highly improbable. Likewise, it might be possible for a woman to have 48DD breasts—even if she is only 5'2" weighing a mere 110 pounds. However, such a women would have breasts that constitute an enormously unlikely amount of the woman's entire body weight. It is more likely that the proportions of these virtual bodies are at least slightly exaggerated (if not altogether fictionalized) in the direction of socio cultural prescriptions for beauty and sexiness—a possibility of which many online participants are acutely aware. As one participant stated, "If they really were 6'2, 185, with 3% bodyfat, and a 8" unit—would they be online trying to pick up a gorilla like me?" Or, as Cleo Odzer (1997:58) describes: "I thought the personlessness [of cybersex]…would cut out the search for the perfect someone of the right age, weight, clothing style, and facial appearance. It doesn't."

It is likely that some participants (perhaps a majority) endow a virtual body with exaggerated physical appearances, sexual abilities, and dimensions of sexual organs. Furthermore, it is probable that these virtual embodiments adhere to (if not extend upon) social and cultural standards of beauty and sexiness. This is certainly true when one considers commonly reported breast sizes. One large online commercial service allows users to create a "profile"—a brief summary of simple demographic and biographic information. The users of the system create their own profiles, and persons can include anything they wish to tell other electronic participants about themselves. Although there is no category for reporting the dimensions of one's virtual body, some people do anyway. A keyword search of member profiles revealed that more than 4,250 persons reported a personal bra size. One out of every three persons listing a bra size (32.7 percent) identified themselves as either D or DD cups.[5] One of two conclusions can be made: either an inordinate number of large-breasted women spend time online, or people who claim to be women tend to exaggerate the breast size of their socially constructed virtual body. The latter conclusion seems far more plausible.

Instead of subverting the "beauty myth" (Wolf 1990), participants "perform" a body that is most often defined in accordance to it. When the body is transformed into a discursive performance without necessary commitment to the physically real, performances become ideal—a reflection of that which is culturally and socially defined as appropriate and desirable. These performances draw meaning from well-established socio cultural scripts for behavior and, therefore, serve to strengthen the "beauty myth," bestowing upon it with greater legitimacy. In this way, the Internet "promotes uniformity more than diversity, homogeneity more than heterogeneity.... Although it has become commonplace to tout the way the Net masks such social markers as race and gender, virtual anonymity does not necessarily lead to relational diversity" (Healy 1996: 62).

In this respect, the power and freedom to define oneself in accordance with cultural standards of beauty is neither a power nor a freedom—it is what Wolf (1990) describes as "the Iron Maiden" of the beauty myth; it confines and encapsulates participants into the tight confines of what is culturally acceptable. The absence of the corporeal body in cybersex only serves to heighten the symbolic importance of the "perfect body." As one respondent explained, "people are playing out a fantasy and the fantasy needs a face and body. Actually, people seem only interested in the body part." Even more to the point, another respondent simply stated, "Don't you know that everyone online is gorgeous?"

These findings fly in the face of those who claim an online egalitarian utopia. Even when entirely disembodied, self-enactments are still subject to the socio cultural constraints imposed upon bodies. As Michael Heim (1991: 74, emphasis in original) suggests, "the stand-in self can never fully *re*present us. The more we mistake the cyberbodies for ourselves, the more the machine twists our selves into the prostheses we are wearing." However, in text cybersex, the prostheses to which Heim refers are not technological; they are instead, cultural. It is worth reminding ourselves that the body is not only an empirical object, it is also a symbolic subject that is presented to others and interpreted according to prevailing systems of socio cultural meaning—even when the body is not physically present in the scene of interaction.

Sex is an act that requires, or is at least dependent upon, physical bodies. One's body in relation to the bodies of others forms the essence of a sexual encounter. Yet, in cyberspace, there can be no physical body or fixed material entity that represents the person. Nonetheless, cybersex does not escape claims of the flesh. Indeed, it fundamentally depends on them, extends upon them, and latently supports cultural and social standards for interpreting them. In text cybersex bodies are transformed into symbols alone—representations, descriptive codes, and words that embody expectations, appearance, and action. This is to say that the body is transformed into a dramaturgical performance. What are being sent to and from computer terminals are not merely words and self-enactments but subject-bodies and body performances. Thus, cybersex is based on claims of the flesh in a discursive embodiment of socio cultural meanings that are attached to a performance and emergent in the interactions between participants.

Conclusions

Because computer-networking technologies allow persons to dislocate selfhood from the corporeal body, one can transform him or herself into someone else. This represents a classic case of vicarious experience— involvement in a role without commitment to that role. In a society where people are expected to be what their role implies, computer-networking technologies have a kind of liberating potential that can be easily seen in online chat environments. As numerous respondents in this chapter indicated, one can be who one might like to be, what one might like to experiment with being, or even be who one does not think one is. This observation is further substantiated by many examples of screen personae that proved untrue. In leisure online communications, persons may not even know something as simple as the actual gender of the person with whom they are communicating. Indeed, self-games abound in the personae playground of online chat environments. However, the findings of this chapter also conclude with equally important evidence contradicting this observation. Body presentations in these online environments are clearly confined within the narrow margins of prevailing cultural norms of beauty and sexual attractiveness. It is apparent that online interaction facilitates both a greater degree of fluidity and greater limitations on the presentation of self than face-to-face interaction. That is, a participant's experiential sense of fluid, open, discursive horizons of multiple potentials for being exists primarily as a freedom to define oneself in accordance with the prevailing standards of what is expected, desired, or mandated by the situation.

Body, Self, and Virtuality

When we tear apart the taken-for-granted seamless surface of reality, as exemplified by online experiences of virtuality, we find a liminal creature existing within the nuts and bolts of the situation (Stone 1995). In online chat and text cybersex, this liminal creature is situated in the boundaries of the experience of disembodiment and hyperfluidity of selfhood. Although the societal imperative is to have one primary persona, that prescription appears firmly affixed to the physical body. In other words, in spite of widely diverse self-enactments, the self-evident, matter of fact, physical just-thereness of the body can comfortably maintain the consistency of personhood in everyday life. As long as one's physical body is present, one can always be certain of one's self—no matter what one is doing. The experience of fluid disembodiment, characteristic of leisure online chat, does not provide this kind of cognitive consistency.

In many ways, this situation encourages the creation of forms of selfhood and body presentations that characterize the postmodern condition. "It is pastiche, a borrowing from diverse imagery, styles, and traditions, including both 'high' and 'low,' mundane and special, and past, present and future, wherever these seem usable: a form of contextless quotation" (Glassner 1990:217). Or, as Trachtenberg (1985:7) describes, it is "performative rather than revelatory, superficial rather than immanent, aleatory rather than systematic, dispersed rather than focused." Yet the fluid and diverse forms by which self and body are presented, negotiated, and validated in online chat and text cybersex do not merely reflect or simply illustrate the postmodern condition. Rather, these conditions represent accommodations to the postmodern condition.

Barry Glassner's (1990:218) analysis of the contemporary fitness movement as "an attempt to reconstruct the self (and in particular the self-body relationship) in a manner that is more felicitous to life in contemporary American culture" bears an uncanny similarity to the findings of this chapter. To Glassner, the heart of the contemporary fitness movement is a "salvation of the self"—"an intimate and holistic marriage between self and body," through being fit, as a means by which "selves are truly embodied." To Glassner, this accommodation allows participants to reconcile the Cartesian twins (mind and body) and resolve principal dualities (i.e. "male-female," "inside-outside," "mortality-immortality"). We may extend upon this and suggest that both the "fit body" and the "virtual body" of online chat text cybersex may be regarded as "a postmodern object par excellence, its image perpetually reconstructed of pieces and colorations added on then discarded" (Glassner 1990: 228). The difference is that the fit "embody the self" while online chat and text cybersex participants *enself the body.* Both accommodations strive to locate selfhood in a safety zone by neutralizing, reducing, or containing meaning (Baudrillard 1980). Both are totalizing— one in the direction of the corporeal, the other in the direction of the symbolic. Yet, in spite of these differences, these two accommodations are fundamentally the same because they both are merely images that are more real than the "real" things they reference (see Glassner 1990). Whereas Glassner suggests that participants in the fitness movement accommodate the postmodern condition by "embodying the self," online chat and text cybersex participants do the same by "enselfing the body." The accommodations are opposite; the effects are nearly identical.

In online chat and text cybersex, participants playfully toy with the virtual actualization of multiple potentials of being. Bodies, selves, and social situations become emergent symbolic elements contingent on communication and interaction. Yet, when people communicate—on the Internet or anywhere else—meaning does not simply arise out of thin air. Answers to questions such as "Who am I?" "What is going on here?" "How shall I understand this other person?" are grounded in a broader socio cultural context. Although disembodied, participants are not apart from socio cultural interpretive structures that provide meaning to self, body, situation, and other. Chat and text cybersex participants fashion a self and body through the same symbolic stock of images that provide meaning in everyday life. Thus, in the dislocated and disembodied context of online interaction, the dramaturgies of the production of a meaningful self and body assume new salience, yet still remain rooted within the same symbolic milieu, utilizing the same sets of resources as any other self and body performance.

What does this indicate about the contemporary relationship between selves and bodies? First, the relationships explored in this chapter are not merely isolated in certain recreational dimensions of cyberspace. Rather, they are embedded within and extensions of much broader shifts in socio-cultural beliefs, practices, and technologies. "These include repeated transgressions of the traditional concept of the body's physical envelope and the locus of human agency" (Stone 1995:16), by virtue of which numerous authors have noted the increasingly pervasive ontological status of multiplicity (see Gergen 1991; Lifton 1993; Stone 1995). In this regard, disembodiment is the embodiment of the experience of multiplicity. That is, if multiple potentials of being have proliferated in technologies of communication, and if the citizens of contemporary culture are indoctrinated with a multitude of selves, then disembodiment provides the ultimate context for multiplicity. This situation allows participants to latently resolve contradictions between a singular corporeal body and experiences of self-multiplicity.

Michel Foucault (1978, 1979) suggests that the emergence of new discourses fundamentally alters the very means by which we understand and interpret body-to-self-to-social-world relationships. If true, then what similar changes might have occurred in conjunction with the emergence of the information era? What new discourses are in operation and in store for the future? Is there any indication of where this is going? There is reason to suggest that contemporary technologies have facilitated the potential for multiple selfhood and opportunities for multiple body manifestations and have resulted in a dissolving of the distinctions between the artificial/technological and natural/biological. Tailor-made bodies, courtesy of cosmetic surgery, are initial elements of this overall phenomenon. The fitness movement, as an attempt to reconcile the self through alterations of the body, is further suggestive of shifts in belief and practice (see Glassner 1990). "We are *already* cyborgs. My mother leads a relative normal life thanks to a pacemaker. Beyond that genetic engineering and nanotechnology...offer us the possibility of literally being able to change our bodies into new and different forms" (Rucker et. al. 1993:100, emphasis in original). Like cosmetic, medical, and fitness alterations, leisure interactions on the Internet extend upon and normalize the potential not only for multiple self-enactments but the malleability of body presentations in a manner that parallels the fluidity of contemporary selfhood. Although Goffman (1959) claims that the body is little more than the "peg" on which we hang a person's self, only in everyday face-to-face reality is the body so inert (Davis 1983); new communication environments challenge these traditional assumptions about the relationships between body, self, and society. What it means to be associated with a body has clearly been altered, and perhaps it will become necessary to completely reconceptualize what it means to be "embodied." Traditional sociological definitions of self as that which is contained or affixed to the body are increasingly questionable assertions—especially with regard to the experiences of virtuality.

Within these techno-social arenas of experience, the meanings of body-self-society relationships manifest themselves in a transformed or transforming state. What emerges is not merely a body-as-container-of-self, or body-as-dramaturgical-prop relationship. Rather, evident in this chapter (and suggestive of broader socio cultural changes) is a body-as-performance relationship. The body is more than a prop that is actively used in a variety of ways to support a multiplicity of self-enactments. The body itself has become multiple, fluid, and discursive. Increasingly, the meanings of the actions define both bodies and selves. As this chapter suggests, the most stable personal characteristic—our sense of where we are in space—is now open to redefinition. There is little doubt that, given the possibilities of selfhood made manifest in the emerging datasphere as a new arena for human experience, and the relationship of these experiences to the bodies that may or may not be grounded in this matrix of virtual experience, new questions arise about what constitutes a person.

NOTES

1. Researchers grossly overlook this fact, many of whom define cybersex as broadly as possible to support the notion that cybersex is common, widespread, and usually a problem (or at least a potential problem with which participants carelessly flirt and which may, at any moment, wreak incalculable disaster on participants' lives). For example, in a recent book on cybersex addiction and recovery (appropriately titled *Cybersex Exposed: Simple Fantasy or Obsession?*), Jennifer Schneider and Robert Weiss (2001:7) couldn't possibly give a broader definition of cybersex: "any form of sexual expression that is accessed through the computer or the Internet." Oddly enough, researchers like Schneider, Weiss, and Al Cooper et al. (1999; 2000) routinely fail to distinguish between *looking at* pornography on the Internet and *participating in* cybersex and thus exploit the ambiguities of the term to support preconceived conclusions about "cybersex compulsivity." This failure to distinguish between looking at pornography and participating in cybersex is perplexing: Do these researchers perceive any significant

difference between looking at traditional print and/or video pornography and having "traditional" face-to-face sex?

2. It must be acknowledged that the relationships between self, body, and society cannot be resolved seamlessly. Distinctions between self and body, body and society, and self and society are fundamentally wed to the basic Cartesian mind-body distinction that permeates the social sciences and is related to the general bifurcation between that which is objective and that which is subjective.

3. In the context of this chapter, I am not concerned with whether cybersex is more like "real" sex or less. In fact, I contend that distinctions between cybersex and "real" sex are far less concrete than often assumed. To a certain extent, even corporeal "conventional" sexual intercourse has always entailed elements of virtual experience. What is the point of sexy lingerie, romantic music, scented candles, and soft-spoken words if not to produce a "virtual" environment for the experience of "real" sexual pleasure?

4. The illusion of stability (i.e. "core self," "personality," etc.) is a social-psychological consequence of stable sets of social relations. If the self is an entity of pure meaning, then, like any system of meanings, it cannot possibly have a singular fixed form.

5. It should be noted that this keyword search only allowed for a maximum of 250 matching entries. Twelve bra sizes produced more than 250 matching entries. This means that there is no way to determine *exactly* how many persons actually report a bra size and the *exact* number of matches for the bra sizes that exceed the maximum 250. However, I feel confident of my interpretation of the data since nine of the twelve bra sizes that exceeded the limits of 250 maximum matches were large C (38–44) or D cups.

❖ CHAPTER FIVE

Televideo Cybersex:
The Naked Self and the Object Body[*]

Although all forms of cybersex involve real-time interaction between participants, there are significant differences between text and televideo cybersex. In text cybersex, participants are restricted to communication in the form of typed words and utterances. For this reason, text cybersex (and online chat in general) represents an ultimate in disembodiment: an extreme "ontological detachment of performed identity from [a] unique and locatable body" (Slater 1998: 96). In this form of interaction, the body is necessarily left behind, at the keyboard, dislocated, and completely absent from the scene of interaction. Yet, because sex requires a body, in text cybersex, participants must evoke them in typed descriptions that emerge interactively with others. In this process, text cybersex participants construct a semiotic virtual body that represents and takes the place of the corporeal body. Because these "virtual bodies" are dislocated from the corporeal, they are not limited by what is empirically real; physical presence does not determine how a person "looks." Instead, appearance depends entirely on information that participants successfully claim for themselves. Thus, participants in text cybersex dramaturgically evoke a body with dimensions and qualities limited only by their knowledge, imagination, and the willingness of others to accept these claims. In this way, text cybersex participants succeed in "enselfing the body." Bodies are transformed into symbols alone—representations, images, and descriptive codes, words of expectation, appearance, and action. The online "virtual body," like the self, becomes an object of pure meaning—a fluid construct that emerges, like the self, as a product of the scene that comes off.

While televideo cybersex also entails considerable text communication, the difference is that one can see the other person(s) involved. By the use of relatively inexpensive digital cameras and client software, participants connect to one another and, in addition to reading what is typed, they may see

---

[*] An early version of the contents of this chapter was published in *Symbolic Interaction* (2002, Volume 25, Number 2, pages 199–227).

each other in live streaming video. Thus, in televideo, "hot chat" is accompanied by the live images of one's partner(s), what they look like, what they are (or are not) wearing, where they are, their moment-by-moment expressions, and what they are doing (often to themselves). For this reason, televideo cybersex is a form of interaction that is nearly diametrical to text cybersex. Televideo cybersex participants embody themselves not in words but in live, moving images that represent them. Indeed, the point of televideo cybersex is to look at images of the bodies of other people and have others look at yours.

Televideo cybersex participants connect with others in computer-mediated environments where they watch each other in live streaming video. In this process they embody themselves, and do so in perhaps the most explicit manner possible: they become a naked sexual object to be looked at. Thus, one cannot, as in the case of text cybersex, simply create a body of unbounded dimensions and attributes. In televideo, the body is not a pure subject of meaning; instead, it is presented as an object—a visible thing, put on display with the intention of being seen, for the purposes of giving and receiving sexual attention and/or arousal.

As we have already learned, given the important role of the body in sex, in cybersex, the presence of the body is not eliminated but only transformed and often magnified. In text cybersex, the body is transformed into an entirely discursive symbolic entity—it exists only as a set of descriptive words that are completely dislocated from the person (hence, an "ultimate in disembodiment," where bodies are "enselfed"). In televideo cybersex, on the other hand, the body is an image to be seen (and hence, selves are embodied). The role of the semiotic "enselfed body" in text-based cybersex has been explored in the previous chapter. This chapter will focus on the role of the body in televideo cybersex. Like the previous chapter, the key question here is: What are relationships among bodies, selves, and social situations within the interactive experience of televideo cybersex, and in what ways do these relationships inform us about the more general experience of selfhood in relationship to the body?

The Object and Subject Body: Sex, Objects, and Sex Objects

> I make myself flesh in order to impel the Other to realize *for herself* and *for me* her own flesh, and my caresses cause my flesh to be born for me insofar as it is for the Other *flesh causing her to be born as flesh*. I make her enjoy my flesh through her flesh in order to compel her to feel herself flesh.
>
> —Jean Paul Sartre, *Being and Nothingness* (1956)

In the previous chapter, it was briefly suggested that the body is always both a subject and an object. What is unique about text cybersex is that the corporeal object-body is dislocated from the scene of interaction; both selves and bodies are fashioned as purely discursive subjects. The situation is not nearly so simple in televideo cybersex. Because televideo cybersex participants can see each other's bodies, it is necessary that we more fully distinguish and detail the relationships between the object-body and subject-body to better understand this experience.

Being treated as an object—a physical body—is a necessary part of human sexuality.[1] In sex, we have little choice in this matter of being an object, but neither do we have this choice in any other form of interaction. The body is always both a noun and a verb; we inhabit an object body (noun) that is subjectively experienced in embodiment (verb). Indeed, the essence of selfhood is to be both object and subject; "the self has the characteristic that it is an object to itself" (Mead 1934: 136), and the same may be said for the body.

While the body may be there as an empirical "thing," its meaning and our ability to conceive of it is no more (or less) innately determined by the qualities of its "thingness" than any other object. In fact, "the body cannot be a direct object to itself. It is an object only to some actor" (Strauss 1993: 110). This "actor" can be an individual or a group of individuals who act toward the body in some fashion (Strauss 1993). They may look at it, admire it, comment on it, examine it, touch it, have sex with it, and so on. It is in a process of interpreting, assigning meaning, and internalizing these interactions that we come to indirectly interpret, know, and understand the body (especially our own).

In this way, as George Herbert Mead (1934: 136) might argue, the body can be an object unto itself, but only insofar as a self is involved; "bodily experiences are for us organized about a self." Surely, the actions of the body are infused with meaning, but it is "the self [that] makes possible the body as meaning" (Gadow 1982: 89); for only by virtue of a self may we see our body as others might. Furthermore, by virtue of this self, we may also act toward our body as others might. We decorate it through clothing, adorn it through accessories, have sex with it through masturbation, and, in extreme duress, even kill it through suicide. In other words, Mead's overall point may be extended: It is not merely that the body and the self are two separate entities; we can only experience either of them indirectly and symbolically by taking the role of the other.[2]

The body and bodily states are experienced indirectly in a process by which corporeal objective and subjective conditions are mediated by shared systems of meaning and thus socially shaped. "Physiological sensations are only the raw material out of which bodily experiences are socially constructed" (Cahill 2001: 47). As analyzed by Mason-Schrock (1996), this relationship between the body, the self, and society could not be more explicitly illustrated than in the case of transsexuals who believe they are born into the wrong body and must look elsewhere for signs of their true gendered selves. But, it would be a mistake to understand this problem as limited to transsexuals and others in similarly extreme conditions. For all of us, the conceptual frameworks by which we interpret the body, the language we use to describe it, the narratives we tell about it, the situations in which we find it, and the definitions we impose on it have an influence on the body as an object and mediate how we experience it as a subject. While we may often apprehend the body in various physiological states—sexual arousal, a drug high (Becker 1963, 1967), pain (Morris 1991), critical illness (Frank 1991), or any other corporeal condition—these are experiences, and like all experiences, they are mediated by social structure, culture, interaction, and social situation.

We may argue that the body is experienced as an object and made meaningful as a subject through the same processes and symbolic capacities by which we acquire a self. As Mead (1934) points out, young infants have no selves, but only acquire them through interaction with others combined with the characteristically human symbolic capacity to take the role of the other with increasing degrees of sophistication. Thus, the capacity for selfhood rests in our ability to see ourselves as others might, and, as Strauss (1993) suggests, the same may be said for the body. Just as we may think about or act toward our self, we may also think about and act toward our body, but to do so requires an ability to interpret, know, and understand our body from the indirect role of others. Just as there is "a dynamic relationship between the acting self (subject) and the viewed self (object)" (Strauss 1993: 112), so too is there a dynamic relationship between the acting body (subject) and the viewed body (object). As Mead might argue, we are born into a body, but we are not born with the capacity to understand that body as an object—to see it as others might, to assess what that body is, what it means, its parts and their relationship to the whole. The body, therefore, comes to acquire meaning in a symbolic process that is no different from any other object.

But surely the body is a special kind of object, if for no other reason than "because it must represent the self in a special sense" (Strauss 1993: 111). It is here, in fact, that the relationships among bodies, selves, and society become entangled and difficult to dissect. Because selfhood is symbolic and cannot be directly observed, it is all too easy to directly equate the self with the body and remain oblivious to distinctions between the two and the broader relationships among bodies, selves, society, and the situations in which we interact. "Embodiment connotes personification, but it also can refer to the body itself as the materialization of otherwise invisible qualities...the body continues to be an omnipresent material mediator of who we are or hope to be" (Holstein and Gubrium 2000: 197), and therefore, this tendency is not entirely without merit.

To be sure, there is an immediate and undeniable relationship between the self and the body. After all, the body and the self are inescapably linked within the total economy of the person. This relationship can be vividly seen in circumstances where the body is severely stigmatized and, thus, negatively affects the self (see Goffman 1963a). More generally, it is obvious that a body is a necessary prerequisite to a self; one can hardly speak of a self that has no body (the loss of a body is tantamount to death; a self without a body is not a person; in common parlance, we often refer to people as bodies when they are dead). As Simmel (1950: 322, 344) suggests, the body is our "first property"; it is that which we unconditionally possess, obeys the will of the owner, and others may categorize as uniquely our own. But still, it would be a mistake to regard the body and the self as synonyms. "Body and self, though inseparable, are not identical" (Gadow 1982: 86). The body does not represent the whole of the self, and the self is not a physiological component of the body. As William James (1892 [1961]: 44, emphasis in original) reminds us:

> ...our bodies themselves, are they simply ours, or are they *us*? Certainly men have been ready to disown their bodies and to regard them as mere vestures, or even as prisons of clay from which they should someday be glad to escape.

While Anselm Strauss (1993: 113, emphasis in original) argues that the distinction between the body and the self "is *only* an analytic artifact," we can afford to be this conceptually casual only if we ignore the important role of the body in relation to the self (see Goffman 1963a; Mason-Schrock 1996) or the degree to which the body can become an independent object and/or subject completely detached from the self within social situations (see Waskul, Douglass, and Edgley 2000). And, of course, to view distinctions between the body and self as mere "analytic artifacts" ignores the increasing prevalence of technologies of social interaction that make it possible to transcend the empirical shell of the body, severing its traditional connection to the self and the situatedness of both in social contexts. More to the point, technologies of communication (especially interaction on the Internet) make it increasingly possible to dislocate selfhood from the body, precariously situating and/or dislocating one or the other from the context of interaction. In some cases, these technologies permit the body to be "enselfed," while in others (such as televideo cybersex) they permit participants to embody the self. Either way, these kinds of separations between self, body, and social context potentially relocate the subject and object of one's actions and inter-actions, and in so doing, they potentially transform and/or expose relation-ships between them.

Nudity and the Net: Voyeurs and Exhibitionists

Televideo cybersex participants meet one another in computer-mediated en-vironments (sometimes one-on-one, sometimes in groups), disrobe (partly or completely, if they are not already naked), display their bodies before others, and comment (usually in typed words) on what they are seeing. This experi-ence typically leads to sexual arousal, self-touching, and masturbation—all of which is usually (but not always) put on display for others to see. Often, these provocative encounters become an emergent "conversation of ges-tures": one participant disrobes and flaunts his or her body; if pleased, the other participant may say so, disrobe, and do the same. One participant may begin to masturbate, inviting the other to do the same. This conversation of gestures continues until its obvious climactic conclusion. Couples (hetero-sexual or homosexual) will sometimes connect with individuals or other couples, in which case, the touching is not always unto themselves. Most often (but not always), these erotic encounters occur between people who do not know one another and whose interaction is merely for the purposes of casual and immediate sexual gratification. Once satisfied, the interaction is typically over, and participants disconnect. However, in spite of these and various other differences, in all cases the eroticism of televideo cybersex is emergent from the intentional displaying of one's body before others in an explicitly sexual manner, being seen while in sexually provocative circum-stances, and watching other people who may be doing the same.[3]

On the surface, televideo cybersex may appear similar to other forms of public or semi-public nudity. Because televideo cybersex participants display their nude body before others, it would seem that the experience would share much in common with nude beaches (J. Douglas 1977), nudist resorts (Weinberg 1965, 1966, 1967), the naturist movement (Bell and Holliday 2000), Finnish saunas (Edelsward 1991), streaking (Toolan, Elkins, Miller, and D'Encarnacao 1974), and the practice of exposing female breasts at

Mardi Gras in New Orleans (Forsyth 1992). Indeed, televideo cybersex is a loose form of semi-public nudity and, as in these other contexts, participants blatantly violate obvious cultural norms (if not the law). In addition, televideo cybersex occurs in an environment that shares a similar liminal quality that situationally suspends the stigma of public nudity. "Clothing modesty is a *ceremony* of everyday life that sustains a nonintimate definition of relationships, and with its voluntary suspension relationships are usually [redefined]" (Weinberg 1966: 21, emphasis in original). Like other contexts of public nudity, televideo cybersex involves "a symbolic separation from the ordinary, a liminal period characterized by separate space, separate time, and separate activities" (Edelsward 1991:193). This separation and liminality, obvious at nude beaches, resorts, saunas, and Mardi Gras, is marked by physical, temporal, normative, and symbolic territories where, in this place, at this time, clothing modesty may be suspended for these reasons. Televideo cybersex is somewhat similar—it occurs in the already liminal world of "cyberspace": a place without space. On the Internet, participants interact from the comfort of their "space" at home (where nudity is permissible), meeting others in a semi-public "place" (where nudity is generally not permissible). The Internet juxtaposes these two very different spaces and places, and therefore, creates a natural environment for liminality: a place separate from one's space where the ordinary norms of everyday life may be easily suspended.

These similarities, however, are quite limited. In final analysis, televideo cybersex differs markedly from other kinds of public nudity (if we can call it "public" at all). Unlike Finnish saunas, nude beaches, resorts, and Mardi Gras, where nudity is purposely antierotic (J. Douglas 1977; Edelsward 1991; Weinberg 1965, 1966, 1967), the nakedness of televideo cybersex is intended to be read sexually. At nude beaches, people "are carefully overcontrolling themselves to avoid any mishaps that would reveal their excitement" (J. Douglas 1977: 79). In these other contexts of public nudity, participants anesthetize any relationship between nakedness and sexuality. Even at Mardi Gras, women who expose their breasts (unflatteringly called "beadwhores") do not do so for sexual purposes (Forsyth 1992). Their exposure is a simple exchange (beads for breasts and vice versa). In fact, with a few exceptions, "beadwhores" at Mardi Gras loathe the opportunism of leering bystanders (Forsyth 1992). At Mardi Gras, like nude beaches and resorts, "voyeurs have become the plague of the nude scene" (J. Douglas 1977: 127). In nude beaches, resorts, Mardi Gras, and Finnish saunas, participants rigidly practice "studied inattention" (J. Douglas 1977: 108), or what Goffman (1963b: 84) would come to call "civil inattention": a circumstance where "one gives to another enough visual notice to demonstrate that one appreciates that the other is present (and that one admits openly to having seen them), while at the next moment withdrawing attention from him so as to express that he does not constitute a target of special curiosity or design."

In contrast, televideo cybersex participants do not desexualize their nudity, they do not practice "civil inattention" toward the naked bodies of others, and they do not loathe the opportunistic gaze of those who may be watching. Instead, they revel in the attention. Unlike naturism, there is no "problematic relationship to sex" (Bell and Holliday 2000); nudity and sex are irrevocably intertwined in televideo cybersex. In televideo cybersex, the nude body is intentionally displayed as an object to be viewed sexually by others, and being seen naked is meant to be arousing. Nudity in televideo cybersex has nothing to do with "a 'philosophy' which is all about bodies in nature" (Bell and Holliday 2000: 127), it is not about defying social norms and values (Toolan et. al. 1974), it is not about the freedom of naturism (J. Douglas 1977), and it does not involve a lifestyle among others who share strong bonds of solidarity (Edelsward 1991; Weinberg 1965, 1966, 1967). In televideo cybersex, being naked is undeniably, frankly, and unambiguously about sex. Stripping and being seen nude are meant to be erotic, and there is no pretense of any other ulterior motives (although, as we shall see, other latent benefits come about).

What is it about stripping, being seen nude, and watching other naked people that makes these experiences so erotic? A common answer is that it appeals to voyeurs ("peepers") and exhibitionists ("flashers") who, due to their deviant sexual tendencies, are gratified by the experience. We must resist these simple answers. It is possible that some televideo cybersex participants may qualify as bona fide voyeurs or exhibitionists. However, distinctions between pathological voyeurs and exhibitionists, as clinically defined by the "tinkering trades," and normative desires to see and be seen naked, are not as clear as they may first appear. Nor does the erotic nudity of televideo cybersex depart from normative experiences of sexuality as substantially as it may first appear.

We can define voyeurism as the practice of observing others in sexually arousing activities or postures. "Most of us are voyeurs to some extent or at some times: we enjoy looking at people, their visual depiction, or reading about sexual activities" (Kupfer 1983: 94). "We all like to see attractive human bodies and will look at them with pleasure even when every part of them is covered except the face or even just a wisp of hair" (J. Douglas 1977: 136). Most of us, including the majority of televideo cybersex participants, are not clinical voyeurs because sexual onlooking makes up a mere part of our sex lives—it is not a substitute for it. Certainly, televideo cybersex participants are often aroused by the voyeurism involved, but the experience is interactive, and because participants can see and respond to one another, there is little gap between what is being watched and how it arouses them. Therefore, the voyeurism of televideo cybersex closely approximates the voyeurism of any sexual experience.

Likewise, we may consider exhibitionism as the practice of putting our body on display as a thing to be looked at for the purposes of sexual arousal or attention. We all are exhibitionists to some extent or at some times: we enjoy being looked at, made to feel desirable, sexy, and attractive. Indeed, a great many of us invest enormous money, time, and effort into diet, exercise, and fashion to obtain this very embodied goal (to say nothing of the ways in which we prune, shave, paint, decorate, and otherwise primp or gussy ourselves up on a routine basis). To be seen and reacted to as a sexual object is, in fact, a normative part of human sexuality. Most of us, including the majority of televideo cybersex participants, are not clinical exhibitionists because being gazed upon makes up a part of our sexual life—it is not a substitute for our physical involvement. The exhibitionism of televideo cybersex closely approximates the exhibitionism of any sexual experience.

If televideo cybersex participants are not clinical voyeurs or exhibitionists, then what accounts for the eroticism of watching others nude and being seen naked? How do televideo participants play with the natural tendencies of voyeurism and exhibitionism in their experiences of cybersex? Because both voyeurism and exhibitionism are directly related to the body (i.e., watching them and having ours be seen), what are the implications and consequences of this experience on the self? Because televideo cybersex participants explicitly display their body as an object for the subjective sexual interpretation of voyeuristic gaze, to what extent may we conclude that televideo cybersex is merely a playful extension of the same social processes by which all bodies are experienced in the interstices of being both object and subject?[4]

Sexuality, Alter-Sexuality, and Cybersex

This chapter examines relationships among bodies, selves, and situated social interactions. Televideo cybersex is used as a strategic lens that allows us to see, better understand, and contemplate these relationships. Even so, it is quite obvious that sexuality is an additional and pertinent theme. Indeed, it would be nearly impossible to analyze relationships between the body, self, and situated social interactions in televideo cybersex without significant reference to sexuality. Thus, while sexuality may not be a direct subject of this chapter, it is essential that it be addressed.

The relationship between sexuality and cybersex is too often assumed, overly simplified, and inadequately considered. Part of the problem is that sexuality itself is an enormously general term that can potentially refer to almost anything that we may regard as "sexual" in relation to the person. Behaviors, attitudes, social customs, fantasies, physical sensations, psychological conditions, sexual orientation—anything that pertains to sex and the person is considered a part of his or her sexuality. Of course, that is precisely how it is most commonly phrased; we speak of "our sexuality," or "their sexuality"; seldom do we consider the possibility that it is merely "a sexuality." That is, in spite of the ambiguities of the term, most often it is assumed that sexuality is connected directly to the person, as something that, although influenced and shaped by others, is within individuals and pertains to their bodies with regard to sex.

Given that computer-mediated communication is a distinctively dislocated and disembodied form of interaction, framing sexuality in this way poses serious problems for the study of sex on the Internet. Even more, we must again remind ourselves that when we push sexuality (or anything else) through the conventions of a medium of communication, a McLuhanesque transformation occurs: What comes out the other side is not the same sexuality, nor is it merely the same sexuality plus or minus one thing or another—the change is not additive or subtractive—it is ecological, becoming something entirely different altogether.

Sex on the Internet constitutes "sexuality as an objectified sphere that is both transgressive and separate from mundane life" (Rival, Slater, and Miller 1998:301). Cybersex participants engage in sexual behaviors in liminal computer-mediated environments, "a place of sexual transgression and 'going beyond': this includes both looking at things forbidden and previously not experienced as well as acting out desires in relation to images or through conversation or fantasy with others" (Rival et. al. 1998: 301). As Laura Rival and her associates indicate (1998: 301), the cardinal attraction to sex on the Internet is the "license simply to float pleasurably through a shamelessly eroticized space." The key point, however, is that

> these pleasures and transgressions evidently depend upon *a clear separation of sexuality from "real life"*: they are without commitment or consequence; the material resources on which they depend (finance, technology, symbolic capital, labour) are obscured from view and experienced as beyond any scarcity...there are no mate-

rial cares or dangers (including disease); no enduring commitments; performance is unproblematic; desire is inexhaustible, as is desirability (everyone is desired and included). Bodies neither fail, nor make non-sexual demands. (Rival et. al. 1998: 301, emphasis added)

For these reasons, it remains unclear whether cybersex participants act out vicarious fantasies that are directly related to their sexuality or merely engage in cybersex as a form of communicative play that is only marginally related to a sexuality (if at all)—it is even more unclear if we can ever tell the difference between the two. On the Internet, a person who is otherwise heterosexual could "go gay" for a brief cybersex encounter for any number of reasons that include the possibility that he or she finds the experience sexually stimulating, but it may also be because he or she finds it amusing or entertaining for other reasons. It is possible that this homosexual cybersex experience, as brief as it may be, is related to the person's sexuality, but it is equally possible that it is not. In fact, Don Slater's (1998: 99) research on Internet Relay Chat sexpic trading found that "most informants were clear that one of the greatest pleasures and attractions of the IRC sexpics scene was not so much the direct indulgence of their own desires as a fascination with the diversity of human sexuality." Slater (1998:106–7) cites at length participants that find "both sexpics and cybersex very boring" and states, "both are merely occasions or opportunities for other pleasures of the scene." Clearly, these kinds of situations raise complex questions, but, at the very least, they are suggestive of the need to conceive of the relationship between sexuality and cybersex differently.

While acknowledging that for some participants cybersex is an expression of "their" sexuality, we must also recognize that cybersex may also involve a form of *alter-sexuality*. We may consider alter-sexuality to refer to sexual experiences that change or become different from those of everyday life; they are a special category of sexual experiences that stand over and against the sexual experiences of everyday life; they are sexual experiences that are bounded within a sphere of experiences that can be comfortably maintained as separate and distinct from everyday life; they are sexual experiences that may not be directly related to the person's sexuality as it is experienced in everyday life; they are kinds of liminal experiences where both intimacy and sexuality may be reinvented in a context of loosened temporal, physical, and normative constraints.

How, then, are we to conceive of this kind of alter-sexuality, especially with respect to the obvious option that cybersex may very well be an expression of a participant's sexuality? Where, then, can we locate sexuality, especially in relation to the dislocated, disembodied, and transgressive experience of cybersex? How may we frame sexuality so it may be dynamic enough to include the enormously varied and rich experiences that are made possible in cybersex? Goffman provides us with a promising approach to framing sexuality and alter-sexuality in relationship to cybersex. To Goffman (1959: 252–253, emphasis in original), the self "is a *product* of a scene that comes off, and it is not a *cause* of it. The self, then, as a performed character, is not an organic thing that has a specific location…it is a dramatic effect arising diffusely from a scene that is presented, and the characteristic issue, the crucial concern, is whether it will be credited or discredited." In identical fashion, we can argue that sexuality is the product of a scene that comes off, and not a cause of it. Sexuality is a dramatic effect arising diffusely from the scene that is presented, and the crucial issue is not whether it is a genuine part of the individual, but whether it is credited or discredited. Framing sexuality in this fashion obliterates the problem of connecting cybersex to the person. As Goffman (1959:252, emphasis in original) might argue, "while this [sexuality] is…*concerning* the individual…[it] does not derive from its possessor, but from the whole scene of his action, being generated by that attribute of local events which renders them interpretable by witnesses."

This chapter proceeds under the assumption that sexuality is a dramaturgical effect; it is a product of a scene that comes off and is not a cause of it. For the purposes of this chapter, sexuality is something that we do; it is not something that we are. Thus, this chapter is interested in how participants do a sexuality in televideo cybersex, with specific attention to how those doings are related to the body, self, and situated social interactions.

Televideo Cybersex: Being a Body and Ephemeral Self-Reduction

Televideo cybersex is a hybrid form of erotica—part pornography and part live erotic entertainment. It is pornography to the extent that the sexually explicit sounds and images that participants see and/or hear are objective and computer-mediated; they appear as digital pictures, videos, and sound bytes that are dislocated from the people they represent. Like pornography, they are something that people look at and listen to. Although the image is "live" (actually, slightly delayed), one does not see other person(s), only images that represent them. Thus, televideo cybersex has much the same voyeuristic appeal as any other kind of pornography with an added twist of being live, fully interactive, and starring honest-to-goodness amateurs:

> It's just for fun. I like looking at other people. More like a game to see if you can guess what they're gonna look like with no clothes on.

What I like most is seeing hot studs!

Of course, I find the whole thing with televideo exciting and men are the reason.... It is exciting from one click to the next to see what is going to pop up in front of you.

I'm just like most males, I get hard just watching a pretty girl putting on a show.

Yet, because televideo cybersex participants are not just watching but also being watched, the experience is more than pornography. Televideo cybersex participants are simultaneously live erotic performers and consumers, casual voyeurs and exhibitionists all at once. Indeed, a central part of the experience of televideo cybersex is to be watched, and therefore, submitting to the gaze of others is usually expected. The general norm of televideo cybersex is, "You can watch me in exchange for me watching you," and, as some participants admit, this norm can be a source of uneasiness. One woman explained, "I've only done it [televideo cybersex] for 2 different guys, but I didn't feel comfortable doing it so now I just flash a booby here or there."

Why would anyone feel compelled to show their body in a circumstance that makes them uncomfortable? Indeed, no one has to do anything on televideo. No one is forced to participate, and persons can simply disconnect at any time for any reason. But even so, all participants seem acutely aware of a very simple reciprocity, or "gift exchange" (Simmel 1950: 392–3), that underlies the experience. Like nude beaches, in televideo cybersex, there is "reciprocity of nude exposure" (J. Douglas 1977: 138). As Simmel (1950: 392) points out, "once we have received something good from another person...we are obliged ethically; we operate under a condition which, though neither social nor legal but moral, is still a coercion...I am caused to return a gift, for instance, by the mere fact that I received it." Thus, it is easy to understand that although one might feel a little embarrassed, uncomfortable, or uneasy about showing oneself naked, this exchange is the price that is paid for the privilege, or "gift," of watching someone else. To the extent that a participant is curious, interested, or aroused by what they are seeing, they may feel compelled to make an equivalent exchange—to return the "gift"— so they may continue to watch. As one man said, "I'd rather watch than show, but people want to see me in return for me seeing them."

Although being seen nude makes some televideo cybersex participants uncomfortable, for the majority this is precisely what makes the experience fun, exciting, and erotic. Many participants are excited by the simple fact that someone wants to see their naked bodies. As one man explained, "knowing that someone wants to watch you is a turn-on." Or, as simply stated by another, televideo cybersex involves a kind of looking-glass eroticism: "being watched makes me feel sexy that someone would want to even watch." Similarly, for other participants, it is the interactivity of being watched by others who comment or otherwise respond to what they are seeing that makes the experience exciting. For example, one participant said, "I prefer to be watched and know I'm watched, with audience participation and involvement." Clearly, for the majority of participants, being seen naked is paramount to the eroticism of televideo cybersex.

The Eros of Stripping, Being Naked, and the Naked Self

Although seemingly obvious, it is essential that we fully understand why stripping and being seen naked can be so erotic. After all, in the absence of touch, that is precisely what televideo cybersex is all about. This is especially important since not all experiences of stripping and being seen naked, even among lovers, are particularly erotic. The obvious answer is that stripping exposes the "body's erotic generators" (Davis 1983: 51)—genitals and other body parts that have been culturally sexualized (buttocks, breasts, legs, and other often fetish-like sexual generators that we normally keep well concealed, or barely and suggestively visible). The mere sight of many of these erotic generators can be very arousing, at least for some people, and for all of us some of the time. Consequently, in non-sexual circumstances where nudity is sometimes required (like medical examinations), it is necessary to evoke dramaturgical tactics to neutralize the power of these erotic generators and create distance between the body and those who are looking at or examining it (see Henslin and Biggs 1971; Smith and Kleinman 1989). Thus, simply exposing one's "erotic generators" for the viewing pleasure of someone else can be erotic for several related reasons. First, it excites those who are watching and thus interactively stimulates our exhibitional imagination. Second, it's the quickest way to turn *any* situation into a sexual one (which is probably a significant reason why nudity is so taboo). Third, it is part of the condition of sex, and therefore, we commonly associate it with sex. And fourth, revealing what is normally concealed is, in and of itself, a rush, a "sneaky thrill" (J. Katz 1988), or a feeling of liberation that comes from blatantly violating a cultural norm.[5]

This, however, is only part of the reason why stripping and being seen naked is an erotic experience. Stripping is also a form of role removal and self-reduction that is an important part of sex. As one man indicated, when nude with others in a sexual situation, one is fully embodied; self and the body are unified in a moment of negotiated sexual pleasure:

Showing yourself in an excited state separates me from the real world. It becomes a focus, no reality, just a sexual moment and a focus on your sexuality. A focus on the sexual organs and also the body seen thru the eyes of others.

"Since most social roles are permanently woven into the clothing worn to play them, whoever undresses casts off these other, sexually irrelevant, social roles—paring himself or herself down to his or her gender role alone...a self now identified with his or her body per se without the usual overlay of extraneous roles that ordinarily dissipate arousal" (Davis 1983: 56). "Just the act of removing one's clothes can help strip away symbolic identity and work roles, allowing one to become merely a body, which is the prerequisite for sexual pleasure" (Baumeister 1991: 38). This last point is essential: Stripping involves a "break through identity boundaries" (Davis 1983: 51), in which we dispose of "all the grand, complex, abstract, wide-ranging definitions of self and become just a body again" (Baumeister 1991: 12). A male participant referred to this identity breakthrough when he said:

I enjoy the freedom of being nude. I have a stressful vocation and it seems the pressures come off with the clothes...Many of us are defined by what we wear. This causes others to pull on you or give you more to do. Such as a doctor's outfit, a nurse's outfit, etc. When naked there is no distinguishing way of identification or anything to link me or anyone else to work. Nudity is a great equalizer. Freedom to just be comes to the front.

As this person indicates, "by taking off our clothes in front of each other, we consciously take off our other selves, our relations to other people, the limits of our relations to each other. We become just a body, outside the normal strictures and plans of daily life. To be naked means to be seen, but below that, to be naked is to be emptied—blank...submerged in the ocean of the moment...a place of ontological surrender.... You lose *you* for a while" (Tisdale 1994: 278–282, emphasis in original). Nakedness sheds some selves, because, in part, it vests others: "sexuality has an essential bodily dimension, and this might well be described as the 'incarnation' or 'submersion' of a person into his body" (Solomon 1974: 338).

It is for this reason that sex in general, and stripping in particular, "must at least have the potential of being one of life's boundary experiences" (Davis 1983: xx). Because stripping and being seen nude break through identity boundaries, many televideo cybersex participants claim to get a lot more than just an orgasm out of the experience. Precipitated in part by a viewing audience, being naked becomes an experiential state that is separate and distinct from the clothed world. Even more, naked is a condition in which one can be that is separate and distinct from the clothed selves of everyday life. Unlike nudity in private, being naked in public and semi-public places is a state in which one is placed as a social object—literally a thing—a body to be seen. Thus, one becomes a "naked self," an objectified nude body that is presented for the purpose of being seen and interpreted as a nude body.

Body-Play: Erotic Looking Glasses
and the Re-enchantment of the Sexual Body

By virtue of this naked self, many televideo cybersex participants are quick to point out not only sexual but also a variety of other personal benefits that come from this experience of being naked. As explained by these two men, being naked has the added benefit of increasing appreciation for and assessment of their bodies:

> When someone is turned on by watching me, it makes me feel that I'm sexier than I truly believe I am...it's nice to get compliments on...the body....I just think it's sexy that people can masturbate and think of me, little ole me.

> Having a few dozen guys tell you how hot you are, etc. really gives you a great outlook on how you see yourself sexually. Positive reinforcement!

For these participants, the excitement that others receive from seeing them nude is repaid to the self by the comforting knowledge that one's body is appealing and desirable. Although a common theme for most participants, this re-enchantment of the sexual body is especially important for those who, for various reasons, claim to be disenchanted with their bodies.

For some participants, this disenchantment has to do with perceptions of their physical appearances, especially regarding age and weight—cultural standards of beauty and sexiness that directly refer to the condition, appearance, and assessment of the corporeal body. In these circumstances, the body may be disappointing to the self, but the sexual attention the individual receives in televideo cybersex serves to undermine that disappointment and re-enchant the sexual body. As one man explained, "Being 48, it makes me feel attractive when someone compliments me on the body." Another man suggested something quite similar: "It feels good when others compliment me. I feel like, even though I am overweight, I am accepted by them." In more detail and complexity, a female participant explained how her weight normally disenchants her feelings of physical attractiveness. She also indicated an acute awareness that many men in televideo cybersex might intentionally manipulate her in order to get what they want (presumably, a sexual performance). But, interestingly enough, her suspicions of deceit are voluntarily suspended. For her, the trade-off is worthwhile because the attention she receives in televideo cybersex serves to re-enchant feelings of sexiness and physical attractiveness:

> [To be seen naked] feels wonderful! Of course, it makes me feel like he desires my body. I know in my head that he is going to say what I want to hear so he can get what he wants, but in turn I get what I want.... As you can see, I'm a pretty good size woman. I'm not uncomfortable about it on here. I feel as desirable as the ladies who are much smaller than me. As a matter of fact, I feel very sexy and seductive on here.

For others, the source of their disenchantment is more ambiguous as they make vague reference to "self-esteem"—a self-imposed assessment of their self of which the body is merely a part. Like the participants cited above, the sexual attention they receive toward their bodies is repaid unto the self, but in this case, the appearance of the body is less significant than the value they attribute to their self. In these conditions, the self is disappointing unto itself, but because the body and the self are inescapably connected, by basking in the positive attention they receive toward their body, they increase the value attributed to the self. This is easy enough to understand—it feels good to be told one is attractive—and that is precisely what this woman indicated:

> I enjoy most the emotional uplift I get from people telling me I am beautiful. I need to feel that I am still attractive.... It's just good to have people tell you you are still attractive especially when I have a low self-esteem.

All of these participants suggested that the way they conceive of their body and self can be disenchanting, and being naked in televideo cybersex re-enchants these internal conceptions of their self and body. However, for some participants, the source of disenchantment had less to do with these kinds of intrapsychic assessments of their body and self, and more to do with interpersonal social relations. In fact, these participants overwhelmingly connected the problem to marriage and long-term partnerships. For some, the problem is related to the routinization of their normal sexual activities and the desire for a different experience: "My main reason for being here is to be sociable, hoping to meet that Mr. Right even though I'm married and have a very active routine sex life at home. The ROUTINE part sucks!" For other participants, the problem is related to the monotonous "mono" of monogamy and the desire for someone different: "I've been in a long-term relationship for almost 5 years, and this way I can remain faithful to that while still getting off with hot guys from around the world." But, for many televideo cybersex participants, it is not the monotonous routinization of monogamous sexual relations that is disenchanting; instead, it is how these relationships intersect with their subjective experiences of body and self.

Murray Davis (1983: 119) argues that "marriage seems almost intentionally designed to make sex boring." We may add that long-term sexual relationships also tend to make our bodies boring. In time, being seen nude by one's lover becomes so commonplace or taken-for-granted that our sexual generators simply run out of gas or otherwise lose their erotic power. As Thomas Nagel (1979: 12–13) reminds us, sexual desire "involves not only desiring another but also...involves a desire that one's partner be aroused by the recognition of one's desire...Desire is therefore not merely the perception of a preexisting embodiment of the other, but ideally a contribution to his further embodiment which in turn enhances the original subject's sense of himself." Thus, it makes sense that when our own nudity no longer generates appreciative erotic power, a classic looking-glass (Cooley 1902 [1964]) process may compel us to feel undesirable, unattractive, inadequate, and thoroughly unsexy. Likewise, it is equally easy to understand why some participants find televideo cybersex helpful in refueling a connection to their corporeal sexual body and then rejuvenating their perceptions of the power of that body to generate eroticism:

> I've been with my wife for so long now that our bodies aren't as exciting as they once were. Our sex life is ok, but without that special quality I sometimes feel like a piece of furniture around the house. When someone is excited about seeing me in all my nudity [in televideo] I suddenly feel sexually awake again.

> Hubby is older than I am. He knows nothing of this. I'm not a complainer but he does not have the passion or the desire, or at least he doesn't know how to show me. ...it's hard for him to tap into the part of my brain that triggers the stimulation. I'm a very erotic person. The feeling of being naughty is a turn-on too. See I'm bad, but I love it.

It is not surprising that these participants claim "therapeutic" value from the experience of interacting with others as a naked sexual object. In fact, this "nude therapy trip" is also a common theme among participants at nude beaches: "a lot of people got into traditional nudism for the nude therapy trip, a kind of commonsense, self-help therapy" (J. Douglas 1977: 158). In addition, the use of nudity as a therapeutic tool in self-help groups, although not a common practice, is documented (see Symonds 1971). At a very general level, to the extent that one attributes "therapeutic value" to the revealing of intimate secrets, being seen nude before others would clearly constitute a dramatic context for such "therapy." More specifically (and less ideologically), being naked before others is a novel condition in which selfhood is shrunk to the body. The normal abstract and multifaceted symbolic layers of selfhood are shed with the clothing in a process by which one becomes a body to be seen. Such a situation begs for interpretation and a redefinition of self that can be harnessed for "therapeutic" purposes, regardless of whether this occurs in the formal context of a self-help group (Symonds 1971) or in more casual experiences such as nude beaches and televideo cybersex. At the very least, as this analysis suggests, certain kinds of body-play may represent important activities in reconstructing relationships between the body as a physical object, and its meaning as a subject, and thus influence the mindful experience of personhood.

For other participants, televideo cybersex is little more than a new and novel mechanism for exploring and cultivating sexuality. For them, nudity is simply a part of the condition of televideo cybersex and the experience of being naked is engulfed by the eroticism of it all. The novelty of the medium and the unique circumstances of being seen and interacting with others in the nude combines to form a potentially powerful erotic experience that builds off of fantasies of promiscuous casual sex, orgies, and similar circumstances in which one's body is offered up as a piece of semi-public property. This is especially evident in the accounts of first-timers and those who have just recently discovered the joys of televideo sex:

> I just tried this on a friends advice and last night I met a guy named ____ ...I mean to be honest, I have never been to that height before, even in actual sex. ...I had several orgasms, no man has done that before.

> Can I tell you a secret? I get more turned on here than I do with my husband.

Regardless of the specific motivations and personal benefits that televideo cybersex participants claim, what all of these people indicate is that being naked in the presence of others is an experience that reduces the whole of the self to the body. It is a totalizing experience where selves are truly embodied. When naked in the presence of others, your awareness shrinks to your body and the immediate present (Baumeister 1991); "the self, abstract and dispersed in everyday reality, becomes embodied and localized" (Davis 1983: 33–34). Thus, televideo cybersex is not a matter of removing the self, rather, it is a condition of shrinking it down to a bare minimum. "The minimum self that a person can have is the body" (Baumeister 1991: 17).

Furthermore, it is not merely that selfhood is shrunk to the body, but also that it is made into an object—a naked sexual body to be looked at and commented on. Clearly, these participants playfully toy with the experience of being an object/subject and, in the process, often gain much more than just sexual gratification. Borrowing heavily from symbolic interactionists' conceptions of the dual prongs of selfhood, in televideo cybersex "the body becomes a focus of interaction and hence a key constituent of the 'me' of the experience" (Glassner 1990: 222). Yet, in these erotic naked episodes, the body is also experienced by way of looking glasses, and thus, creates a space "where the 'I' does stand a chance, where one can both participate in and respond to the informational overbearance of the body" (Glassner 1990: 223). Translating these terms into more common vocabulary: in televideo cybersex, the tension between the acting body and interpreting self (what interactionists call the "I") and the viewed person/body (what they call the "me") functions as a kind of personal ritual of self-renewal (Davis 1983): by manipulating the relationships between them, one may temporarily enfeeble the self by being a body in order to return to selfhood reaffirmed.

Just Another Body: The Face, Self, and Body

The junction between being a naked body and watching one is central to the experience of televideo cybersex. These bodies, however, have selves. Although selfhood may be shrunk to a bare minimum (a body), that "shrinking" is only temporary, ephemeral, and squarely situated in these liminal online environments. In everyday life, we are rarely able to so clearly separate the body from the self, and therefore, in everyday life, we are far more accountable for what we do with, to, and by our bodies. It should be no surprise, then, that in "real life," these participants rarely (if ever) indulge in this kind of explicit and promiscuous sex-play. As one woman reported, "I would not dare flirt like this in public. The men seem more open and flirtatious as well." A male participant expressed the same sentiment: "There is no way in hell that I'd do anything like this in real life!"

Indeed, being a body is more than just a by-product of nudity in these semi-public online places. It is also a prerequisite for preserving the integrity and dignity of the self. Although long-term friendships and intimate "real-life" relationships can and do sometimes emerge from televideo encounters, most of the time the experience is fleeting, anonymous, and casual. Being seen naked (and often masturbating) in anonymous and casual semi-public social situations is, after all, quite taboo, and therefore, as Jack Douglas (1977: 58) reported with regard to nude beaches, many people are "afraid of being seen by friends or associates and branded a weirdo or a sexual pervert."

So as not to implicate the self in these promiscuous virtual encounters, great efforts are usually taken to keep identity detached from the body. Because televideo cybersex participants can control the angle and zoom of the digital camera, this delicate dramaturgical task is most commonly achieved by simply not showing one's face. As one man vividly explained:

> It's all very anonymous, at least for me. I rarely show my face, therefore I am just another body. Once I show my face, it becomes too personal—I'm not just a body anymore!

Televideo cybersex participants clearly manipulate a fascinating relationship between the face, the body, and the self. As televideo cybersex participants have discovered, a body without a face is a (no)body. That is, a body without a face is a body without a self—it is "just another body."[6] Televideo cybersex participants are not the only ones to have learned this. Parts Models, a New York agency that specializes in modeling body parts, does not supply erotic models, but when asked to find a female fifty to sixty years old, willing to expose her nude chest for a health article in *Self* magazine, the editor claimed: "I screened the magazine very carefully and once they assured me they wouldn't use the model's face it was no problem" (Norwich 1987: 51). David Roos of Gilla-Roos, Ltd, an agency that also specializes in parts modeling, described a similar circumstance: "we just shot a commercial for an airline. They wanted a 'chubby man who would stand in the water. Naked.' Of course we found their man...but only because everything would be exposed except his face" (Norwich 1987: 53).

Clearly, the face occupies a supreme position in connecting or disconnecting the self with the body. After all, one's face is the most identifiable feature of one's body and self—it is the equivalent of an identity fingerprint. The face is the single physiological human feature that concretely conjoins the corporeal body with the self. The face assumes paramount importance in identifying us as a person who is distinct from anyone else. Although the face is just another part of the body, we tend to regard it as more uniquely ours than any other part. While we might be quite fond of other body parts, the rest of our body generally does not have this special signifying quality.

It is significant to note that the face is the one part of the body that is almost always seen naked and, therefore, it is not only constantly accessible as a means of identification, it is also extremely vulnerable. As one woman indicated, because our face is so uniquely ours and represents the most critical identifier of who we are, some fear that if they were to show it, someone might recognize them: "I am afraid to show my face. I live in a small town and someone could notice me." This is reiterated by another man: "I do show my face, but not when I am getting to know someone.... I am well known in my community and also around the state where I live. You never know who might be on-line and checking in." Indeed, as this man makes perfectly clear, the vulnerability of "showing face" makes cybersex—the safest of sex— potentially risky:

> I don't show my face at first. I started doing that after an embarrassing situation with a coworker not long after installing the camera and software.... I had been putting on a show for someone with no cam for several days. A guy at work finally told me it was him [and he] wanted to play in real time. It was very embarrassing.... Since then, I've just felt more secure not showing face. [I] still do show face, but only after I'm sure I don't know the other person.[7]

Of great concern to many televideo cybersex participants is the fact that often these images can be saved (hence, reproduced and distributed). Once again, since televideo cybersex is a hybrid form of erotica—part pornography and part live entertainment—these images can be collected and potentially used to incriminate the self. Acutely aware of the possibility for extreme embarrassment, televideo cybersex participants keep their face (and identity) conspicuously absent from images of their body as a tactic for managing these anxieties. These two men were particularly concerned about the obvious fact that one cannot control what others might do with the images of them and managed those anxieties by concealing their face:

> I don't show my face because [the image] can be saved.

> I don't show my face just to protect my privacy. If this could not be saved, I would show my face more.

By not "showing face," the greater and most significant portion of one's identity is concealed, making the process of being a body experientially complete. Indeed, by the frequent practice of detaching the face from the body, televideo cybersex environments often become a virtual meat locker of neck-down naked people—a mass of headless torsos and genitals. As crudely remarked by Lawrence Langner (1991:92), "groups of men and women without clothing appear to lose their individuality and become merely a herd, like any other herd of animals." While we may appreciate Langner's point, it is also apparent that he has not seen too many naked bodies. After all, upon closer inspection, no two bodies are alike (in fact, one of the more interesting things about televideo cybersex environments is the education one receives in just how different bodies and body parts really are). But still, Langner has a point: without a self or identity attached to them, they are just penises, breasts, buttocks, and torsos—each slightly different, but essentially the same and just as good as the next. Thus, by not showing one's face, the experience also becomes exceedingly impersonal.

Some participants dislike the impersonal element of televideo cybersex; others thrive on it. As one man explained, those who like the impersonality refer to the need to be a "thing" from time to time:

> Yes, it's impersonal and therefore not very fulfilling. But sometimes being impersonal is exactly what I crave! To simply be seen, looked at, desired, and for someone to comment on what they are seeing.

Those who dislike the impersonal virtual meat locker of televideo cybersex refer to what almost everyone already knows: sex (even on the Internet) is much more fun when it's with a person rather than an object. This would seem obvious, but still, it often remains grossly overlooked. For example, cultural critic Mark Dery claims "the only thing better than making love *like* a machine, it seems, is making love *with* a machine" (1992: 42–43, emphasis in original). Dery is wrong. No matter how technologically advanced, expensive, or what gee-whizzery it promises, almost anyone would prefer to have sex with a person rather than a machine or object. Machines and objects are no match for people in sexual affection. Machines and objects cannot and will not replace people as objects of sexual desire. Almost everyone would rather have sex with a person because machines do not (and cannot) have a self. And that is precisely what some televideo cybersex participants expressed:

> Most of the time, I personally do not continue to carry on conversations with men who will not show me their face, with a few exceptions. lol However it does become very impersonal when all they want from you is to help them masturbate, without ever seeing their face or expressions. I have connected before and just watched without feeling any sense of connection at all. [In these situations it was] very impersonal, and I wondered if the satisfaction they felt was comforting at all.

> It's a bit impersonal. When I get to know the other person more, I show face.

> My face is my best feature. How can people tell that you are happy doing what you do online if they can't see your expressions? The eyes, they are not alone as gate-

ways to the soul, all language of the body shows the souls words and wishes....
[Showing your face] lets people into your world a little more...to let them look into
more of my world through my face and my expressions, brings them closer to the
reason that I am masturbating online—not because I'm horny, or want attention, but
because it makes me feel good to share the experience.

While nudity transforms one's self and body into a unified object, nudity
in the context of these semi-public televideo environments makes the objectification even more extreme. Part of this is a function of the nature and limitations of the medium. Although televideo images are "live" and interactive,
they are still computer-mediated and thus dislocated from the people who
then display themselves within. To have sex with someone is to touch and
feel another body—sensations that are not a part of televideo cybersex.
Televideo lacks these qualities, and therefore, as aptly described by one man,
televideo cybersex is "like looking at a live magazine."

In another way, being a body is also a function of the social context. Social taboos against nudity and masturbation in public and semi-public places
assure that participants will be prone to conceal their identity and present
themselves as only a body or body part—a detached penis, vagina, or pair of
breasts with no face, identity, or self. On one level, being a body acts back
unto the self in an erotic "looking-glass" process of ritual self-renewal. On
another level, when one presents oneself as a body with no self, everything
that makes a person sexy, desirable, and interesting is stripped away, and
having sex with a selfless body is not much different from having sex with
any other object.

The Gendered Body in Televideo Cybersex

Although "you can watch me if I can watch you" may be the norm of televideo cybersex, in practice the sexual exchange is rarely this egalitarian.
Televideo cybersex participants may reduce the self to the body, but, as we
all know, not all bodies are equal. Without doubt, in televideo, some people
by virtue of their bodies have more power to call the sexual shots, and many
can command a performance without having to perform themselves. In
televideo cybersex, this power is undeniably available to women.[8] Furthermore, although all women have this power, the more attractive they are the
more sexual power they may wield (if they wish).

Because there are substantially fewer women than men in televideo
cybersex environments, men compete with other men to attract the attention
of women, and this competition can get fierce. As one man casually
remarked, "The fight for attractive people is awful." A key way that men
attempt to attract women is through conspicuously advertising themselves by

attract women is through conspicuously advertising themselves by their screen names and, when additional fields are available, providing descriptions of their physical features and special talents. Thus, screen names and identifying fields become spaces for men to boast about themselves. Consider, for example, the following listings (all of which have been modified to protect participants' anonymity):

HardWood	25 year old STUD	6', 185, Muscles, HUNG
BadMAN	Fit, BIG COCK	Cum watch me j/o
BoyinNylon	Very Submissive	Spank me!
HungDude	8" plus a lil' more	Cute and playful!

Women also use screen names and identifying fields. However, women are less likely to emphasize their physical features and instead tend to advertise their preferences so as to attract the right applicant for the job. Consider, for example, the following listings (all of which have been modified to protect participants' anonymity):

BeHARD	Want HOT men	I like BIG dicks!
Bi_ANN	Prefer petite	Tempt Me—NO MEN!
Drippingpussy	4 BIG cock	Must be CUT!

Attracting the initial attention of a woman is only part of the problem men face in securing a televideo cybersex encounter. Because of these competitive conditions, when a woman shows interest in a man, he must also be prepared to please her, listen, obey, and follow her instructions. With so many men to choose from, women do not need to tolerate someone they find rude, offensive, unimaginative, ugly, uncooperative, or otherwise inadequate. In addition, televideo cybersex is computer-mediated. Participants interact within the liminal ether of electronic space. For better and worse, they cannot touch each other, nor do they have any idea where in the world they are located. Thus, the distance afforded by the computer-mediated context adds an additional element of safety for all participants (not just women). Not surprisingly, many women feel a sense of liberation that is related to this combination of distance, control, and what they perceive to be a reversal of sexual power:[9]

It's sometimes empowering cuz I know I have the power to block someone I don't like or whatever.

I think it's liberating. I have more power to choose and flirt with men without worrying about physical repercussions to me. It is functional.

Being watched is erotic, and if they make me feel like a piece of meat, I have control.

You choose who you want to see you, so you have a good idea how it will turn out.

Like other aspects of televideo cybersex, this privilege and power comes at a cost. Unfortunately, the cost of this power ends up making televideo cybersex more like the everyday micropolitics of sex than a departure from them. Because men are competing with other men for the attention of women, some men become exceptionally boorish, brazen, rude, and often just plain offensive. Consequently, women sometimes find the heightened attention excessive. As one woman explained, "I think it is harassing sometimes the way they are so bold to just say or show something without knowing who or what is on the other end!" Another participant made the situation especially clear: "I think it's appalling. How worse could you treat someone than as a slab of fuck-flesh to be beaten off for?" One woman went so far as to categorize men on televideo according to the degree to which they are an annoyance with respect to their behaviors. Her account is particularly instructive of the kind of excesses that women commonly encounter in these environments:

I kinda got an order of which I perceive these men who are clamoring for attention: 1. Normal: Fully dressed, at work or home, for chat or looking for love. 2. Semi's: without shirt with a big grin, usually married waiting to play with someone who is willing. 3. Regulars: showing nothing but their penis with an erection waiting for an orgasm (some are there every day) some say they have 2 or 3 orgasms a day. Wow. They must never have real sex. lol 4. Kinkys: either doms, subs, bis, each are wanting something out of the ordinary in order for them to become excited. 5. Pervs: these are some sick mf's. They do ungodly things with their bodies, want to do really sick things—example, shove a Pepsi can up his ass (I couldn't believe it), have sex with animals, etc. Overall it is very exciting when you are dealing with types one, two and sometimes three. lol But when it comes to the last two they harass the shit out of you, bothering you the whole time...until you finally have to just put them on your ignore list.

The naked body is not freed from social strictures. At the very least, the naked body is, in glaring clarity, a gendered body.[10] Many social roles can be shed along with the clothing worn to play them, but some roles are permanently fashioned into our skin, and the more of that skin we show, the more salient that role becomes. While Naomi McCormick and John Leonard (1996:110) claim that in cyberspace "we are freed from our physical bodies," that freedom is no more or less possible to obtain in televideo environments than in everyday life. In fact, it should be no surprise that the traditional micropolitics of sex become paramount—even exaggerated—when genders interact in the nude (on the Internet, or anywhere else).

Conclusions

Computer-mediated communication is a unique form of social interaction that can be characterized by several related conditions. Most importantly, computer-mediated communication is a *dislocated* and *disembodied* form of interaction. It occurs in a social "place" without necessary connection to geographic "space," where the activities of participants and experiences of self are not necessarily contained within or affixed to corporeal bodies. At the very least, these characteristics create a context in which online interaction may assume unique forms that have the potential to challenge traditional understandings of self, body, social situations, and the relationships among them.

In spite of various qualities of the medium that characterize this unique form of interaction, it is misleading and inaccurate to assume that these qualities spontaneously result in certain experiential conditions. For example, as the previous chapter has explored, characteristics of computer-mediated communication may be related to experiences of "ultimate disembodiment." Yet, on the other hand, as this chapter has illustrated, the same characteristics of the medium may also allow for experiences of "total embodiment"—a condition in which one is embodied within an image of one's own body.

As will be discussed shortly, this distinction between "ultimate disembodiment" in text cybersex and "total embodiment" in televideo cybersex is almost entirely conceptual; in both text and televideo cybersex, the body is a virtual object. The distinction I have made between "ultimate disembodiment" and "total embodiment" is analytical, not experiential; in either text or televideo cybersex, a participant may be embodied or disembodied by the experience. I have merely used these concepts to better frame and explore certain dimensions of personhood on the Internet. The important point is not that one form of online interaction automatically results in one form of social psychological experience. It is not the medium alone that causes certain communicative and experiential outcomes. Instead, in the final analysis, computer-mediated communication is like all other forms of human interaction: it is emergent from the purposes of the communication, within situated contexts, that is negotiated with others. Although important, the medium is just one of many variables that intersect with the experience of online inter-action.

Televideo Cybersex and the Profanation of the Sacred Sexual Body

In any scene of interaction, the body is a subject that is experienced but also an object that is viewed and acted upon by others. How that body appears is of paramount significance to the definition of the situation, the interactions that transpire, and the scene that comes off. Consequently, most of us most of the time are quite conscious of how others may view our bodies, and while we may sometimes intentionally manage that image, at the very least we al-ways make certain that we have some clothing on when in public or poten-tially in the public view. Failure to do so would be disastrous for reasons that run deep into the intersection of society, self, and the acculturated body.

At a very young age, we are socialized to society's proscriptions and prescriptions concerning the situations, circumstances, and purposes of al-lowable and forbidden genital exposure (Henslin and Biggs 1971). These dictates are deep, powerful, and compelling—violating them can result in extreme embarrassment, and stigma, and make one subject to intense social control. "The iron rule against public nudity is one of the most basic, most important rules in our society. It is so important, so rigidly adhered to that, once it is learned at approximately the age of three, it is almost never spoken again" (J. Douglas 1977: 9). In no uncertain terms, we learn to distinguish the publicly viewed body from "private parts" that we keep carefully con-cealed and the dramaturgical conventions necessary to protect the latter from being made into the former. These mandates are so strong and clearly de-fined that we can state with utmost certainty that there are only three general categories of persons that can legitimately approach the naked body of someone else: authorized sex partners, medical practitioners, and the parents of very young children (who, significantly enough, have a limited capacity for selfhood). It is also worth noting that all three must be very careful about how they approach the naked body: sex partners must be careful lest they be guilty of rape or abuse, medical practitioners make certain to use elaborate dramaturgical tactics to assure that the context is desexualized and, if possi-ble, the person is detached from the body completely (see Henslin and Biggs 1971; Smith and Kleinman 1989), parents must be careful lest they be guilty of sexual misconduct (at the very least, they must recognize that after a cer-tain age they are no longer privy to seeing their children's naked bodies). Even in common circumstances where we must modestly disrobe, like public bathrooms, there are elaborate dramaturgical rituals and carefully designed barriers that are intended to provide physical and normative shields against potential audiences (see Cahill et. al. 1985). There is little doubt that "because of the sexual significance of nudity there are elaborate cultural normative and value systems surrounding the naked body and the degree to, and circumstances in, which it may be exposed to others" (Bryant 1982: 27). Thus, there can also be little doubt that the treatment of and deference to the naked body is deeply ritualistic, and by adhering to these rituals, the naked body achieves a sacred status.[11]

Beneath the facades of clothing is a sacred sexual body; compartmentalized segments of flesh that we may touch only in privacy, share only with the closest of intimates, are protected by stringent rules, and are filled to the brink with profound meaning to the individual, those who may see it, and culture. Under normal circumstances, access to the sacred naked body of others means power and privilege and, therefore, connotes carefully detailed rights and responsibilities. Indeed, the naked body can be conceived of as a sacred object: it is surrounded by rules that protect the object from being profaned, rituals that dictate who may (and may not) approach the sacred, under what conditions, and what may (and may not) be done in these circumstances (Durkheim 1965 [1912]). As Goffman (1971) and Durkheim might suggest, it is adherence to these rituals that maintains the "sacred" sexual body as an object of ultimate value, and while threats of profanation may occur occasionally, the sacredness of the naked body can be comfortably maintained so long as we abide by the rules.

In this way, televideo cybersex participants are cultural heretics. They blatantly profane the sacred, and should their heresy be made public, they would likely suffer extreme embarrassment and stigma. Televideo cybersex participants display their naked and often fully aroused sexual body as an image refracted through the ether of electronic space, in semi-public places, dispersed to any geographic location, for anyone to see (so long as one has access to the televideo environment). Televideo cybersex participants reveal what we normally keep well concealed and exhibit behaviors in semi-public places that we normally keep quite private. A large part of the eroticism and feelings of liberation that comes from televideo cybersex stems precisely from this cultural heresy and the breaking of these rules. Televideo cybersex participants intentionally display their naked bodies as profane objects, a piece of semi-public property that, like a community park, may be looked at or used by anyone who happens to be passing by.

It is possible that this kind of dramatic and extreme revealing of secrets is linked to a culture awash with revelation. Indeed, "soul-baring has become a national obsession" (Edgley and Brissett 1999: 5). "The Victorian era—the height of the print culture—was a time of 'secrets'.... Our own age, in contrast, is fascinated by exposure. Indeed, the act of exposure itself now seems to excite us more than the content of the secrets exposed" (Meyrowitz 1985: 311). However, unlike this kind of pitiful and shameless exposure, because selfhood can be detached from the body in these computer-mediated environments (and hence identity may be concealed), televideo cybersex participants may reap the benefits and remain relatively safe from the potential consequences of revealing of what is normally secret and situationally profaning what is normally sacred. By revealing their naked body in these semi-public spaces, televideo cybersex participants assure a context of interaction that will be intensely personal, yet by concealing their identity, the self is suspended as a spectator in these erotic episodes. Thus, on one hand, the benefits are the same for televideo cybersex participants as they are for fitness enthusiasts, "an intimate and holistic marriage between self and body" (see Glassner 1990: 221). On the other hand, in order to achieve this, televideo cybersex participants reduce the whole of the self to the body in a process that profanes the sacred and assures that the usual complex and multidimensional self is made into a mere object, an image to be seen, and, therefore, not a person at all. Although televideo cybersex is far more ephemeral, it is a condition, like fitness, where selfhood is simplified by directly equating it to the body.

Reading the Object/Subject Body:
Virtual and Corporeal Bodies, Everyday Life, and the Internet

Text cybersex participants clearly create a semiotic "virtual body" through typed communication with others in online environments, resulting in an "enselfment" of the body. In contrast, televideo cybersex participants display images of their corporeal bodies, the experience of which serves to embody the self. As this chapter has illustrated, these experiences can be seen as diametrical. However, in spite of these critical differences, in final analysis, the body is a virtual object in both contexts. While this may be clearly seen in the semiotic virtual bodies of text cybersex (where the whole of the body is emergent in typed communication, with no connection to the corporeal), this is no less evident in televideo cybersex. Instead of words and typed descriptions, the virtual body of televideo cybersex is an image, something that one looks at, and nothing more. It is not a corporeal body that one may touch or otherwise physically interact with but a virtual body to which one is a spectator.

As this chapter has emphasized, televideo cybersex participants make their body an object for spectators in erotic episodes and, in a looking-glass process, those spectators influence how the self conceives of itself and the body. This process is made explicit by virtue of the technology that allows participants to exchange images that represent their bodies. In fact, both the corporeal body and the self are dislocated from the image that represents them, making it especially easy for participants to conceive of themselves as an object—to fully objectify themselves in the experience of televideo cybersex. Furthermore, participants can see and respond to the images of their own bodies and, thus, act toward, manage, and interpret that image as others might. At the very least, participants must remain aware of and manipulate the images of their body in order to prevent "showing face" and to assure the appropriate camera angle, zoom, and focus necessary to perform the "conversation of gestures" inherent in the mutual masturbation of televideo cybersex. It is certain that televideo cybersex participants are acutely aware of and manage the virtual body that represents them in these erotic encounters, and consequently, they too become spectators unto themselves.

It should not be surprising that televideo cybersex participants claim personal and even therapeutic benefits from this experience. The technology, social context, and deeply personal ("sacred") nature of what is happening in these encounters assures that they will see and respond to their bodies represented in images and interactively negotiate the meaning of these events within a context of others. All of this shapes not only the experience of televideo cybersex, but also stands to impact how participants conceive of their selves and bodies. However, this process is not fundamentally different from the ways in which the body manifests itself in everyday life; it is simply an exaggeration of it. In any scene of interaction the body will be placed as a social object and we will most often manipulate its appearance—this is essential to the overall definition of the situation and the self we claim within it. Through clothing, accessories, and cosmetics, we manipulate what and how others perceive of the corporeal body in everyday life, and in so doing, the body in everyday life assumes a virtual ontological status that is not that much different from the images sent and received by televideo cybersex participants.

Consequently, this chapter has illustrated one of many important relationships among the body, self, and the situated contexts in which both are experienced. In televideo cybersex, participants become an object in a pro-

cess by which selfhood is reduced to a naked gendered body. The usual multiplicity of selves and social roles that define a person are cast off with the
clothing so that one may become a thing to be observed, appreciated, interacted with, and perhaps even harassed. Of course, these bodies have selves,
evidenced by the identity concealment tactics that participants use to protect
themselves from the potential embarrassment that could result from the discovery of these promiscuous online sexual escapades. But, even more importantly, as the body becomes an object to be seen by others, it also becomes an
object for the self, and this process is no different on the Internet than it is in
everyday life. In a classic "looking glass" (Cooley 1902 [1964]), the body is
symbolically reflected back onto the self via other participants within these
televideo contexts, and this process, it impacts how the body is conceived,
appreciated, and apprehended by the self.

What all of this suggests is that it is possible to understand the body as
an object and subject that is distinct from the self. This is especially evident
in sex (and made even more explicit in televideo cybersex). However, having
said this, it is also essential to note that it would be a mistake to analytically
sever the connection between selfhood and the body. In truth, the boundaries
between the self, the body, and the situated contexts in which both are located will never be clearly demarcated, and any attempt to do so would seriously lack credibility. A better approach is to understand the varied dynamics
of the relationships between bodies, selves, and the situated contexts in
which both are located. If we were to agree that "the essence of human existence is embodiment, that the self is inseparable from the body...[than] the
problem of the relation between self and the body is not solved; it only becomes more interesting" (Gadow 1982: 86). Sometimes the body is nothing
more than a necessary but relatively unimportant "peg on which something
of collaborative manufacture will be hung for a time" (Goffman 1959: 252–
253). In other circumstances, the body and its appearance assume paramount
importance in defining the situation and interaction. Sometimes, selfhood is
only marginally related to the body while at other times the whole of the self
is shrunk to the body. Sometimes, we have no choice in the inevitable union
of the body and the self while at other times, we may intentionally manipulate the relationships between them in order to sever the connection between
the self and the body. In a culture seemingly obsessed with bodies (their appearance, health, fitness, etc.), it would be a serious mistake to ignore these
and the myriad of other relationships among the body, the acting individual,
and the social worlds in which both are located.

What this chapter suggests is not a need to analytically sever the connections among bodies, selves, and social situations, but to further explore the relationships among them. It is not that self and body are separate and distinct entities, but rather that one does not automatically imply the other, and neither are simply innate things that the individual has. Surely the body, unlike the self, is an empirical entity that can be observed, measured, and seen to exist in time and space independent of our own conscious awareness of it. But even so, what that body means is no more (or less) innately predetermined, nor is it any more (or less) socially shaped than any other object. As the participants in this chapter have vividly illustrated, the body is not just a thing that exists, it is something that is read, interpreted, presented, concealed, and made meaningful in an ongoing negotiated process of communication and situated social interactions.

NOTES

1. While many feminists legitimately argue that patriarchal social and cultural structures reduce women to mere objects, one can hardly deny that being an object is essential to sexual experience. This is not to dismiss the criticism of feminists, but only to suggest that there is a fundamental difference between being *made* into an object and choosing to *be* an object. The former is domination; the latter is a normal if not "joyous part of sexual life" (Nussbaum 1995: 279). The issue here is really power and hegemony, not sexual objectification per se. In sex, we willingly become a corporeal object for the experience of pleasure in our partner(s) and ourselves. To be made into an object with no regard to our consent is to be dominated in other domains.

2. It would seem that Mead (1934: 136) was quite aware of this point:

   It is perfectly true that the eye can see the foot, but it does not see the body as a whole. We cannot see our backs; we can feel certain portions of them, if we are agile, but we cannot get an experience of our whole body.... The mere ability to experience different parts of the body is not different from the experience of a table. ...The body does not experience itself as a whole....

3. This chapter is only concerned with televideo cybersex—casual and usually anonymous televideo sexual encounters between participants who do it for fun. Certainly, televideo has been used widely for the pursuit of profit within the Internet pornography industry (see Lane 2000). "Live" webcam performances are a relatively common and apparently lucrative service for the entrepreneurs of virtual pornography. However, these for-profit webcam services differ from televideo cybersex in that performers are paid to put on a show, the audience pays to watch, and the performers rarely (if ever) see their customers. Those that pay the fee can "peep" in on the show and sometimes communicate with the performer through typed text, but there is little interactivity and the motives of the performer(s) are at least partly financial. This chapter focuses on people who use televideo

for sex, for fun, and for free (not considering ISP fees, cost of the digital camera, and cost of the client software) with other participants who are doing the same.

4. I should note that while some use it for sex, televideo is also used for a wide variety of mundane purposes—it would be a mistake to consider televideo as merely a technology of compu-perversion. Perhaps the most common use of this technology is innocently personal—to see and interact with distant relatives and friends. Other non-erotic uses involve corporate applications by cities, schools, businesses, and a variety of other organizations that employ televideo to display live images on the World Wide Web to anyone anywhere who may wish to watch whatever might be going on in these places. There are various "scientific uses" (for those interested in peeping in on things related to science), including exotic research outpost cams (including images from AASTO, the US South Pole station), microscope cams (live images of various microbiological life forms including bacteria and cancer cells), and space cams (images from high-power telescopes). Indeed, because it is relatively inexpensive and easy to implement, it seems that everyone and everything has a webcam these days—homes, airports, gardens, beaches, cafés, coffee shops, buildings, parks, lakes, bridges, roads, shopping malls, museums, mountains, volcanoes, resorts, coffee pots, vending machines, women, men, infants, couples, and even pets. Entire websites are devoted to listing these various webcams in all of these categories, and many more.

5. "Social theorists have resisted strongly the recognition that deviance is not merely a reaction against something negative in a person's background but a reaching for exquisite possibilities...deviant persons also appreciate the economy of doing evil for characterizing the self generally: it is literally wonderful. Through being deviant for a moment, the person may portray his or her general, if usually hidden, charismatic potential" (J. Katz 1988: 73).

6. A fascinating exception appears in *Le viol* (1934), a painting by the Belgian surrealist Rene Magritte. In this surreal anatomical rearrangement, a woman's face and body are one and the same. It is a portrait where the outline of a face is the outline of a woman's torso wrapped in hair, eyes are represented by a pair of shapely breasts, the nose by a bellybutton, the mouth is a patch of pubic hair. Not surprisingly, this painting has been subject to substantial feminist criticism for the manner in which it "fragments the female by turning her into a sexual body" (Gubar 1989: 49). Some have argued that the female face is erased by the female torso, making her sightless, senseless, and dumb, an abject and obscene artistic rendition of female submission and servitude (Gubar 1989). According to Gracyk (1991: 125), it is "one of the most graphic subordinations of a woman to an object, for it reduces her to genitalia." What is fascinating about this image (and the controversy surrounding it) is that few seem to ponder whether Magritte intended to (super)impose a body on a face, a face on a body, or both. To some critics, the image offensively objectifies women by blatantly equating a face with a body. Yet, it can also be suggested that the image subjectifies the body by equating it directly to the face. And, it might further be suggested that it does both simultaneously. When critics look at *Le viol,* they see one dimension of Magritte's surreal and philosophical riddle, seeing only a woman's body in place of a face. It is interesting that they do not see a face in place of a

body, or both at once. Even if Magritte had not intended it to be so, *Le viol* impressively mines the fascinating relationship between bodies and faces—the object and subject of embodiment—revealing (at the very least) the taken-for-granted cultural tendency to see the image as one thing and not another.

7. It is significant to note that televideo cybersex participants reverse the traditional expressive order of the body that is normally experienced in the ritual of courtship. Traditionally, lovers become familiar with one another's faces first and genitals last. As these participants indicated, in televideo cybersex, genitals come first and faces are seen last (if at all).

8. Obviously, this section refers only to heterosexual one-on-one televideo cybersex. I am certain (but cannot be sure) that the relationship is quite different among couples that connect with other couples, gay men, and gay women. In each category, it is very likely that the distribution of sexual power will assume different forms in relation to the body and the context of interaction. I focus here on one-on-one heterosexual encounters for several reasons. First, these are the most common sexual relationships in televideo cybersex. Second, a full analysis of sexual preference in relation to televideo cybersex is not necessary for my overall argument—this one component is sufficient. Third, and most significantly, I had no difficulty collecting data from gay men, but gay women and couples were not at all interested in talking to me, and therefore, I have little or no data on them. It is quite apparent that gay women and couples are all too familiar with the deceptive tactics of some unscrupulous heterosexual men who often try to cop a sneaky peek, or overtly con them into a brief encounter. Such tactics are not that uncommon. One participant in this study gloated when she told me how she uses pictures in a magazine (combined with a slight blurring on the focus of the cam) to temporarily con people into thinking she is something she is not. Of course, the picture does not move and, therefore, the con is short-lived, but deceitful tactics such as these have made gay women and couples very suspicious, and almost impossible for this man to interview.

9. This perception of feminine sexual liberation and power in televideo cybersex is actually far more complicated than suggested here, and is experienced as real for reasons that are far more legitimate than this analysis suggests. Although I am skeptical about much of what passes for "sexual freedom" in cybersex, it is still worth reminding ourselves that we remain a society where women continue to be punished or otherwise found guilty for actively seeking sexual pleasure and overtly acting on sexual desire, as opposed to simply "following their hearts" (L. Williams 1989: 209, 259–60). Moreover, at least some of the controversy over contemporary high-tech, easy-access, in-the-comfort-of-your-home pornography is attributable to the fact that it is now brought not only into locations where children might wander, but also "into the domain of traditional female space" (L. Williams 1989: 283). There is little doubt that sex and pornography on the Internet poses unprecedented opportunity for women to indulge sexual interests and desires in a context that is relatively safe, and to do so in a rather carefree fashion that men apparently take for granted. Where else do woman find "license simply to float pleasurably through a shamelessly eroticized space" (Rival et. al. 1998: 301)? Where else do woman find equivalent opportunities to explore sexual interests and desires without paying the price

of stigma, punishment, shame, or other fees charged courtesy of the sexual double standard? In short, the analysis presented here is critical of the extent of sexual freedom given the form in which it appears on the Internet, but even so, I have to acknowledge that the freedoms expressed by the women in cybersex are quite real and legitimate (albeit for reasons that are beyond this analysis). The mere fact that woman can "float pleasurably through shamelessly eroticized space" on the Internet is in itself a major development—even if the forms of those sexual expressions adhere to what appears to be the same old androcentric sexuality.

10   A similar analysis could be done on the role of race and age in the body politics of the nude in televideo cybersex. Just as one's gender cannot be removed by clothing, neither can race nor age (although age can be faked to some extent). Indeed, many televideo cybersex environments display relatively common sexual fantasies that build off race and age stereotypes. Particularly evident are fantasies of the experienced (and often dominant) elder mistress with her boy-toy, and fantasies that build on the mythical girth of the African penis.

11   This is oversimplified. Both the naked body and the clothed body may be considered "sacred" (in the Durkheimian and Goffmanian sense). The naked body achieves sacred status as a result of the rules that govern it and the deference with which it is treated. But, the deference and rules regarding the treatment of the naked body exist and persist because clothing is "one of the most basic and pervasive symbols of civilization" (J. Douglas 1977: 39–40). It is basic and pervasive because it is in part through clothing that we physically and symbolically transcend ourselves. Or, in other words, clothing successfully fails to cover the body; clothing is a partial synthetic exterior skin with the delightful capacity to accentuate some parts of the body while concealing others. Although we come into the world "covered in a skin…man soon discovered that he could make himself not only one additional skin, but practically as many as he liked, an endless variety of them, to meet his every need and fancy" (Langner 1991: 4–5). Of course, these "skins" are produced within the milieu of society, culture, and institutional contexts, in accordance to formal and informal social mandates. Consequently, "adornment creates a highly specific synthesis of the great convergent and divergent forces of the individual and society" (Simmel 1950: 344). That is, clothing is a key mechanism by which we simultaneously express our individuality while also declaring our union with others (Simmel 1904 [1971]). Thus, in clothing, we literally wear society, as both a convergence with society and divergent expression of our unique self. It is, therefore, not unreasonable to suggest that it is through clothing (in part) that we are no longer akin to the animal world, but to the world of gods, spirits, and society (Langner 1991). For these reasons, the source of sacredness of the naked body lies in its ever-present potential to subvert society itself; the naked body is anti-sacred because it is the clothed body that is sacred to social order.

❖ CHAPTER SIX

Transparency and Transformation: The Internet, Personhood, and Everyday Life

Some people are surprised to learn that I seldom talk about my research in the courses I teach. Years of scholarship on chat and cybersex remain subjects that I write about, but rarely discuss in classrooms (or anywhere else for that matter). Granted, I will occasionally recite anecdotal research tales to illustrate concepts or ideas, but I seldom specify the kind of work that has consumed my scholarly interests since 1994. As a student, I endured far too many courses that, although called something else, were really about the professor's research interests—I refuse to commit the same injustice against my own students. My scholarly interests are just that, *my* scholarly interests; they are directly related to the courses I teach, but I have not the ego to presume that they are synonymous with them. Even so, the number of students who discover my work routinely surprises me. Once "discovered," I'm flattered to learn that students are often interested in what I have written, frequently asking encouraging, penetrating, and well-considered questions. Admittedly, I take pleasure in these moments.

Given the kind of research that I have done, I have always enjoyed certain privileges of being considered "contemporary": aware, knowledgeable, and interested in newly emerging technologies of social interaction; intent on understanding their influence on the people we are, or are becoming. In addition, I am both blessed and cursed with the most boyish of looks; I am not only younger than most of my university colleagues, but also appear considerably younger than I actually am. For better and worse, I can pass for an incoming freshman at any college or university—more than once I have had to show identification to suspicious university staff who did not believe my claims of faculty status. It is, perhaps, because of both these reasons that students and colleagues alike have seemed inclined to regard my work, and me, as one coworker once described, as "closer to the cutting edge of cool." Frankly, I don't know what to make of any of this, but I can say I was not entirely aware of it until very recently.

In the fall of 2001, I was teaching a course composed entirely of freshman students. Not a single student in the class was over the age of twenty. At one point in the course, as occasionally happens, a student inquired about my research on online chat and cybersex, and, like many times before, I proceeded to explain. But this time, something curious happened. Normally when I discuss my research I am able to sense a kind of air of electrified interest and curiosity; typically there are numerous questions and related discussions. This time, however, none of these things were apparent. That day, I stood in a classroom discussing my research while thirty-five sets of eyes looked upon me, not with interest or curiosity, but with confusion and a kind of perplexed amusement. It was an uncomfortable and unfamiliar experience. It dawned on me, then and there, that these students could not see what I considered so unique and novel about the Internet; they were boggled by why I would spend years studying something so mundane. Furthermore, what I had to say was nowhere near the "cutting edge of cool" but, instead, seen as just another example of the usual academic musings about things that everyone else already claims to know and understand. When I asked, they told me: they were introduced to the Internet in grade school. And then it occurred to me, something I should have seen coming for a long time: for these new college students, the Internet is not only nothing new—it never was; it has, for all practical purposes, always been a part of their life.

I will forever remember that moment in class. It was the first of what I'm sure will be many times I will feel "old." My purposes in telling this tale, however, are not to whine about my own inevitable maturation. Instead, I recall this incident because it is indicative of an important shift in the broader social and cultural environment in which people apprehend the Internet; it is suggestive of a change in how we interact with the Internet. Throughout the 1990s, we could legitimately seek to understand the Internet as a new, emergent, and explosive communication technology. By the turn of the century, none of these things are nearly so apparent anymore. For the most part, the Internet has already become just another part of our mundane world, neither new nor unique but instead another one of many utterly transparent technologies of communication and social interaction, different from, but no more significant than, television, telephones, cell phones, or facsimiles, one of many means by which information can be gathered, correspondence can be made, items can be purchased, and entertainment can be obtained. For these reasons, more so than any other, it is necessary to shift our approaches to understanding the Internet and its relationship to the experience of everyday life in society.

We have entered what may be considered a third era of the Internet. The first era owes much to the cold war. I'm sure others have noted how a paranoid cold war communication structure would ultimately come to be coopted by consumerism and commercial capitalism. I will not belabor the postmodern ironies of it all here. What is important about this first era is that the Internet was conceived and its skeleton constructed. The second era was the "wiring" phase that occurred through the 1990s. In the second era, the social and cultural world got progressively "wired," and that wiring brought with it innumerable changes, accompanied by a host of hopes and fears about what this amazing new technology would do to and for us. This proved to be a time ripe for utopian dreams and distopian fears, as the technology seemed to embody all that we hoped and feared for society and ourselves. While few, if any, of these utopian and distopian imaginings have been realized, for a moment, the Internet dared us to dream—for better and worse.[1] Now, things have changed again. We are all wired up; the hopes and fears have dissipated; books about the social and cultural significance of the Internet are no longer best-sellers; the loud hype has been reduced to increasingly infrequent claims—and even those tend to originate in corporate ad campaigns that are, all too clearly, intended only to influence spending habits. There is a strange silence, a calm acquiescence. We have entered a third era—an era of transparency. No longer novel, no longer new, the Internet is part of the world taken for granted—we think it neither unusual nor amazing how the Internet brings us into contact with people, places, and information in various ways. The Internet is a routine part of life in a privileged society (or, at least, privileged life within society). In fact, the only things we find unusual or amazing are the occasional quaint people we encounter who are *not* connected.

I have no interest in making futuristic predictions, but even so, it is difficult to avoid concluding that this current era of the Internet—this era of transparency—promises to last. It won't last indefinitely, but it seems fair to suggest that it will remain until the advent of another unpredictable communicative innovation that culminates in a similar kind of paradigm shift. Or, in other words, we shall have to wait until another medium for human communication and interaction is introduced into our communicative universe. Meanwhile, the technologies of the Internet will continue to change, there will also be new uses and fads, and scholars will still struggle to make sense of it all. But the Internet itself will continue to go the way of other technologies of communication. In this respect, the Internet will become increasingly like television. Indeed, as my students taught me one autumn day in 2001, this is already happening.

Television, for example, is something we all watch but few see. That is, television has become transparent. One of the most amazing things about television is that, even though we all watch it, we do not really look at it; instead, we look right through it. When watching television we see what is on, the medium itself is invisible. In fact, the television itself is invisible; when watching we seldom notice we are looking at depthless moving images, framed in plastic, on a cubed screen.[2] Even more, television programs change from season to season, new fads come and go, but television itself has not changed much since the advent of color. And yet, all of the most important effects of television have come, arguably, because we do not see it and because it has not changed much. In similar fashion, the Internet is rapidly becoming just as transparent as television. For this reason, we would be wise to assume that all of the most important transformative effects are still to come; the changes we witnessed throughout the 1990s were a mere trailer for the feature presentation, only in this show, the changes are likely to happen in subtle and silent ways that we shall hardly notice. Indeed, these changes may be happening right now; we are simply unable to fully perceive them.

Few will deny that mediums of communication can have a substantial influence on social and cultural worlds, sometimes significantly altering the experience of life in society. Skeptics need only examine the history of television and consider its radical influence on politics, education, consumer activity, family, work, fashion, and innumerable other dimensions of everyday life. However, this transformative power does not reside solely within the medium. Instead, it is emergent from how we interact with that communication technology. These interactions can be quite complex, but our ability (or inability) to perceive of the medium is one central and extremely important feature that will considerably influence its transformative potential.

It is difficult for a communication medium to exert significant influence on who we are, what we know, and how we know these things to be true when we clearly see the medium itself. Likewise, a communication medium will exert its most profound influence when we do not (or cannot) see it. After all, other social processes share this characteristic. For example, this same relationship can be seen in domination and oppression: it is difficult to effectively dominate and oppress people who know they are being coerced. Although still dominated and still oppressed, so long as people know it is happening, they can resist, if even in the smallest of ways. It is much easier and far more efficient to subordinate people who are unaware they are being manipulated. One cannot resist that of which one is oblivious. Domination and oppression are complete and total—a totalitarian's dream—when invisible, taken-for-granted, and routine parts of everyday life.

I do not wish to imply that communication mediums are mechanisms of oppression (although some have). However, I do wish to emphasize that, like oppression, the greater and most significant influences will begin to occur just as soon as they are imperceptible. My point is hardly that radical: when the Internet is something new, it is also something different. For both of these reasons, we not only see the medium itself, we also apprehend it as something decidedly separate and distinct from our normative everyday experiences of life in society. When it is no longer new and no longer different, then these distinctions vanish; it becomes an extension of our everyday experiences of life in society. Things previously considered different now become normalized. Unique and novel experiences become mundane. When this happens, then—and perhaps only then—can the Internet truly transform us.

For this reason, it is obvious that all of the most important work on the Internet, and what it has to tell us about the kind of people we are, is yet to be done. It is still much to early too discern the nature of that work; the Internet is not completely translucent yet. But one thing does seem clear: it will become increasingly more difficult to see through the mundane transparency of the Internet while, at the same time, our ability to do so is most crucial. Thus, we must begin to shift our perspectives; it is necessary that we sharpen our conceptual and analytical focus, entertain new ideas, and carefully assess our approach to understanding with a renewed spirit of critical review.

Throughout this text I have sought to offer analysis and discussion of online chat and cybersex that illustrate what these experiences can tell us about personhood in contemporary society. Some of these illustrations are timeless, while others seem to capture more contemporary struggles for meaning in an uncertain, rapidly changing, and ambiguous (post)modern society. I will conclude by summarizing and commenting on what I believe are the three most important general insights that can be gleaned from these studies. Each, I believe, may be especially promising as analytical starting points for those who wish to see through the increasingly mundane transparency of the Internet, and begin to assess its emergent relationship to the experience of life in society.

First, while it is obvious that online chat and cybersex reveal distinct forms of computer-mediated interaction that differ from everyday experiences of life in society, they often do so in ways that are conspicuously familiar. Although unique in many important respects, experiences of self, body, and social interaction on the Internet are built up from the same processes by which these things are experienced in everyday life. It has been my contention that a more complete understanding of personhood on the Internet requires that we not only appreciate how these experiences are different, but also how they are similar to commonplace everyday life—the latter of which is too often overlooked. It has also been my contention that approaching human experiences on the Internet in this way holds a special promise of illuminating more generic properties, characteristics, qualities, and processes of personhood on and off-line. In short, I have suggested and sought to illustrate how we can use the Internet as a lens for better seeing, contemplating, and understanding the experience of life in society. I sincerely believe that this kind of approach to understanding personhood on the Internet remains the most hopeful framework, not so much for understanding the Internet, but for the more important project of better understanding ourselves. For example:

By understanding how online chat and cybersex participants discursively write themselves into existence, we gain insight into the ways that we all communicate to others the kind of people we are. Whether on the Internet or elsewhere, we are always engaged in a process of discursive and non-discursive communication by which selfhood is interactively crafted. Online chat and cybersex are not departures, but extensions of this basic process.

By understanding how online chat and cybersex participants playfully toy with multiple representations of who they are, we gain clarity on the ways that we all manage multiple representations of our multifaceted selves. In our contemporary world—which some reasonably call postmodern—it is clear that we are all multiples. As it turns out, we all can provide a plurality of answers to the question "Who am I?" each answer depending largely on where we are at, what we are doing, and with whom. Selfhood is altered and divided—sometimes significantly and sometimes subtly—between work, home, school, church, during leisure activities, when in the presence of colleagues, family, friends, or strangers. Never before has it been so instructive to remind ourselves that the word "person" comes from the Greek *persona,* referring to an actor's face mask. In our everyday experiences of life in society, it has become increasingly obvious that to be a person is to be a wearer of masks. Online chat is not a departure from, but an extension of, this quality.

By understanding how online chat participants negotiate the "cultural crack" between experiences of self-multiplicity and the cultural prerogative for a unitary self, we gain clues to how we all negotiate these discrepancies in our lives. While the works of cognitive theorists do not especially move me, still, it may be worth considering the extent to which contemporary social and cultural conditions have become increasingly disjointed from the ways in which we understand ourselves in this world, resulting in a collective form of cognitive dissonance. If Leon Festinger (1957) is even marginally correct—that this kind of inconsistency or dissonance makes people uneasy enough to alter either their ideas or activities to make them more consonant with each other—then "cognitive dissonance" is no longer an aberration, but a normative part of our everyday struggles to make meaning in this world. Regardless of whether we call it a "cultural crack," "cultural lag," or a collective form of "cognitive dissonance," the point is that, increasingly, the experience of life in society does not match social and cultural understandings of that experience. By examining the ways that online chat participants resolve this "dissonance," we are glimpsing into the heart of important changes in the frameworks by which we all come to understand what makes us "us," me "me," you "you," and what does (and does not) distinguish between these categories. Online chat is not a departure from this compelling struggle for meaning; instead, it is a microcosm that magnifies them.

By understanding how text cybersex participants construct fluid semiotic subject-bodies, we also learn about how we all craft the appearances of our bodies that, like in text cybersex, are subjects to be read by others. Everywhere and in all circumstances, the body is never just an object—a corporeal thing—it is also (and more importantly) a subject to be seen, interpreted, and acted upon. We all know this, and actively manipulate how others "read" our bodies in ways that are not much different from text cybersex participants. Like text cybersex participants, we habitually draw from a vast repertoire of social and cultural meanings to make our bodies an embodiment of who we are, or think we are. It is certainly true that in online chat or text cybersex, a woman may be a man, a man a woman, an adult a child, and so on. However, these body-games do not depart from everyday life quite so much as they might first appear to; instead, they only reveal the subjective ways in which we all *do* our bodies in everyday life: how men make themselves into men, women into women, adults into adults, and children into children. In the end, RuPaul had it right: "Honey, you're born naked. Everything else is drag."

By understanding how televideo cybersex participants manipulate relationships between the body as an object and the body as a subject in a "looking-glass" process that acts back influencing how the self conceives of itself and the body, we see with great clarity a very complex relationship that is identical to how we achieve an embodied sense of personhood in everyday life. All of us are a body; all of us are more than a body. We all craft a self in situated social interaction, pinched between the body and the "more than body." For all of us, the body is an object and a subject; so too is the self both an object and a subject. It is out of the immense complexities of the relationships between these qualities of personhood, as they are experienced in social situations, that we achieve a unique sense of individual personhood. Televideo cybersex is not a departure from these complicated relationships; instead, it extends upon them in ways that make them more easily seen.

In these and many other ways, human experiences on the Internet tell us volumes about the nature of personhood in contemporary society. Certainly online chat and cybersex participants are not always conscious of the broader social and cultural implications of their Internet activities. But even so, perhaps never before have so many people been so actively engaged in social psychological experiments that are so distinctively at the cutting edge of important historical, social, and cultural transformations. No one has more succinctly summarized the significance of these activities than Rosanne Allucqure Stone (1995) who once neatly referred to this as "reality hacking." We have yet to take full advantage of these unprecedented and utterly amazing techno-socio-cultural experiences of "reality hacking" to glean important insights about what kind of people we are (or are becoming).

The second major analytical position that has appeared repeatedly throughout this analysis is that communication and interaction on the Internet are *dislocated*. Communication and interaction on the Internet dislocate the meaning and significance of time, space, and the corporeal body to social interaction and the experience of personhood. I have, admittedly, focused primarily on one dimension of this dislocation—the ways in which online communication and interaction dislocate selfhood from the corporeal body. To a certain extent, it is fair to say that I've been both fascinated and obsessed with this experience of disembodiment and what it can tell us about relationships between self, body, and situated social interaction in general.

What I find truly unique about online interaction and cybersex is that the corporeal body is located in one place, and a symbolic representation of it is located elsewhere. This, in itself, is not without precedence (see Flowers 1998), but the quintessence of online communication and interaction (especially cybersex)—the remarkable interaction between the corporeal body, its virtual representation, and the situated communicative context—surely is. In reference to erotic CD-ROM "games," Linda Williams (1989: 312 emphasis in original) describes the body in these virtual environments as "felt to be in two places at once...pleasure consists in a body that is uncannily both *here...and there...*there is a sustained simultaneous dividedness of attention and blurring of the distinction between the virtual bodies on screen—one of which is now presumed to be 'my' own—and my own 'carnal density' here where I sit before the screen." As I have sought to analyze, this body in online chat and cybersex—precariously perched in "two places at once"—is not only a unique characteristic of real-time computer-mediated communication, it is also what makes these experiences ideal for the kind of social psychological analysis that can fundamentally inform and transform the ways in which we think about bodies, selves, situated social interaction, and the nature of personhood, because it makes opaque relationships that are normally quite transparent.

Finally, it is so obvious that it is embarrassing to even mention it: communications and interactions on the Internet are mediated by computer technologies. Of course, this lends these forms of social interaction distinct characteristics. Yet, as I have already sought to explain, this does not mean that everything about computer-mediated communication is new and novel; it is not a complete departure from everyday experiences of life in society. To a certain extent, when we look at and think about computer-mediated communications, we tend to see obvious glaring differences that blind us to how these experiences are similar to communication and interaction in everyday life. It is exceptionally easy to see the ways that computer-mediated communications depart from everyday experiences of social interaction; it is more difficult yet far more promising, illuminating, and important to explore characteristics that these experiences share in common. I have already made a case for this argument, but I have not yet pushed this idea far enough.

If we think most generally about experiences of communication, social interaction, and personhood on the Internet and compare them with the same experiences in everyday life, there are many differences and similarities that we may identify, depending on our point of view. There is one issue, however, that increasingly requires our utmost consideration. Until recently we have quite comfortably maintained a distinction between face-to-face and mediated forms of communication, often considering the latter the poor bastard cousins of the former. Face-to-face social interaction is often regarded as the paramount, paradigmatic, and ultimate foundation from which all the other various mediated forms of "para-social" (Horton and Wohl 1956) communication are but artificial and synthetic substitutes. Peter Berger and Thomas Luckmann (1966: 28) have gone so far as to claim "the most important experience of others takes place in the face-to-face situation, which is the prototypical case of social interaction. All other cases are derivatives of it." Face-to-face is real, genuine, natural, and true; mediated forms of communication are, well, "mediated," and hence, made artificial by the ways in which technology supplants authentic interaction. It is becoming increasingly clear, however, that the only thing false and artificial is this distinction itself: all forms of communication and interaction are mediated.

If there is anything real, true, genuine, or natural about human communication and interaction it is that, always and everywhere, it is mediated. Sometimes, it is overtly mediated by apparent technologies that are all too easy to see (print, radio, telephones, television, the Internet, etc.). What is often unseen, however, is that even face-to-face interaction is mediated, only by language, spoken words, voice intonation, eye contact, posture, contextual variables, social roles, statuses, clothing, gestures, and all the various other symbols by which meaning is conveyed through the medium of face-to-face interaction. There is no such thing as unmediated human communication. Whatever it means to engage in unmediated communication and social interaction, it is beyond my capacity to imagine. Even if we omit everything else, communication and interaction necessarily involve language, which, in-and-of-itself, mediates, structures, and organizes the world according to its own symbolic code. Few have articulated this more lucidly than B.L. Whorf:

> We dissect nature along lines laid down by our native language. The categories and types we isolate form the world of phenomena we do not find there because they stare every observer in the face; on the contrary, the world is present in a kaleidoscopic flux of impressions which has to be organized by our minds—and this means largely by the linguistics system in our minds. We cut nature up, organize it into concepts, and ascribe significance as we do largely because we are parties to an agreement to organize it in this way.

Even if the distinction between mediated and non-mediated communication is legitimate (and I don't think it is), one can hardly deny that the relevance of that distinction has rapidly eroded. At the very least, there has been an undeniable "shrinking of the differences between live and mediated encounters" (Meyrowitz 1985: 121). Regardless, in the end, it matters little whether our conceptual categories were flawed or if human activities have changed so much as to require new definitions; the point is that this distinction fails to assist us in better understanding ourselves. It is increasingly essential that we grasp the meaning and implications of what scholars like Joshua Meyrowitz (1985: 122) have sought to tell us for some time: "face-to-face interaction is no longer the only determinant of personal and intimate interaction...electronic media are unique in that they mask the difference between direct and indirect communication."

All forms of communication and interaction are mediated; the important question is not whether communication and interaction are mediated, but by what means? Marshall McLuhan was right; every medium of communication and interaction contains its own language, grammar, and syntax that exerts a profound influence on what that medium is used to produce. What we must add to this is that face-to-face interaction has become one of many mediums by which people act, interact, and communicate with others. No longer is it reasonable to presume that face-to-face communication holds a monopoly on the paramount, one-and-true, genuine form of social interaction. Perhaps, in the end, we will come to learn that, in their transparency, mediums of communication—whether television, the Internet, telephones, or whatever—end up making opaque the arbitrary distinction between "mediated" and face-to-face interaction, revealing their most important characteristic, the one thing that is so bluntly human about them: their capacity for communicative expression, matched only by our endless capacity to delight in that expression.

NOTES

1. What is perhaps most remarkable and disappointing about these dreams and hopes is that we invested them into technology. It would seem that we have arrived at a sad and cynical point in our cultural history where we no longer dare to dream about what we can do collectively to make our world a better place. Often, it would seem, we have given up on these kinds of aspirations, regarding them as mere utopian fantasies. But, the dreams themselves have not gone away. Instead, we invest into our technologies that which we once dared to believe we could do for ourselves. Indeed, it would seem that our faith in technology has no limits while faith in our selves seems to have reached new depths of despair. No one, for example, ever seemed to question whether we could put a remote control car on the surface of Mars, creating what is surely the most expensive remote control car in the history of the world. While we knew it might take some time and trial, most people were quite confident that such a task was well within our potential. But when asked if we can reduce poverty, create global peace, feed the hungry, provide health care for all who are in need, and so on, I am utterly stunned at the number of people who feel these things are beyond our power and potential to either create or control. Similarly, for a moment in the 1990s, we dared to dream that the Internet would bring about a more democratic society, more egalitarian relationships, and other improvements in our social and cultural world. Oddly enough, few seemed to consider that such societies cannot be made by technology—it requires something of *us*; it requires a realization of a certain potential within our collective way of life. Perhaps, in the end, I am nothing more than an Enlightenment idealist that is, on one hand, encouraged by how technology dares us to dream, but on the other, deeply disappointed by how we have come to embody technology with characteristics that only we can do for ourselves. I would like to

believe, however, that one day we might be able to see how these hopes are misplaced and find renewed faith in our own collective potential.

2. The transparency of television can be easily observed in everyday life; it does not require tedious supporting evidence. It is instructive enough to note that the greatest contributions to our understandings of television come from scholars and theorists who have sought to make visible a medium that has become transparent to our perception (McLuhan being, perhaps, the best example). But even so, it is significant to note that the transparency of television can also be observed by merely looking at the structure and design of television sets through time. Consider, for example, console television sets. These large, and sometimes aesthetically pleasing, console television sets were not simply designed to be looked *through*—they were something to be looked *at*. Unlike contemporary television sets, these were items of furniture. The same observation can be made for radio and record players—the former have even become collectors' items, not because they are good radios, but because old upright radios are often beautiful pieces of furniture. Even more, for many years in my youth I worked for an auction company that sold estates. I often encountered console televisions and radio/record players. I vividly recall, not only their sometimes-enormous size and weight, but how these were used in people's homes. They were used for much more than merely watching television, listening to the radio, or playing records. They were almost always adorned with live or plastic plants, doilies, and various knickknacks—sometimes even books. It is difficult to avoid noticing how impossible it would be to use a contemporary television in the same way—one must purchase an "entertainment center" for such purposes; the contemporary television is not a piece of furniture. In fact, the contemporary television is so aesthetically displeasing that those who can afford it will find ways to hide its appearance when not in use—when forced to see the television itself we seem to realize its ugliness and seek to camouflage or conceal it. The difference here is stark: for people to whom the medium was not transparent, the television was an aesthetically pleasing multipurpose item of furniture—they could not only see the television set, it was designed to be seen. For people to whom the medium is transparent, the television is an eyesore whose appearance is tolerated only because it is consumed by its sole function of *performance*—otherwise it is something that ought not be seen.

❖ APPENDIX

Methods and Data: Doing Ethnographic Research on the Internet

Beginning in the mid 1990s and continuing to the present, researchers in all fields of the social sciences have steadily recognized the scholarly potential of the Internet. While still marginalized to a certain degree—sometimes still considered a mere trend or fad—the Internet has, nonetheless, attracted increased interest and legitimacy as a valuable context for social and cultural research. However, doing this kind of research presents new methodological problems. In this short appendix, I will briefly identify and discuss some of these issues. I will begin by generally discussing critical distinctions that researchers must consider in order to effectively frame Internet studies and will further discuss specific methods and problems I encountered while collecting and analyzing data for the studies included in this book (for even more details see Waskul 2002; Waskul and Douglass 1996, 1997; Waskul, Douglass, and Edgley 2000).

Framing Internet Studies

In all the ways that matter the most, quality empirical research on the Internet requires careful attention to all the same variables that researchers would consider in any other context. Topping the list of things to consider, effective ethnographic research on the Internet—like anywhere else—necessitates that researcher(s) clearly understand what they are studying. This would seem so obvious as not to require mention, but in the case of the Internet, the situation is not always so simple.

It is essential to understand that the Internet is not one single thing, technology, medium, or environment for human interaction. Instead, we can legitimately consider it a *meta*-medium, *meta*-technology, and *meta*-environment—a medium of mediums, a technology of technologies, and a place of places. Understanding this also makes clear it is a social environment of social environments, a loosely connected network of vast and varied forms of human association, communication, and community. For this reason alone, effective and appropriate methods for conducting research on the Internet will be heavily influenced by the nature, structure, and organization of what the researcher is studying—if the researcher is unclear about what he or she is studying, the entire project could be easily compromised.

This kind of problem is not unique to the Internet; it is endemic to social research in general. But even so, for Internet studies, the situation is further complicated. Because the Internet is necessarily a mediated context for communication and social interaction, it is easy to lose sight or mistake the frame of one's analysis; it is easy to intend upon studying one thing only to end up collecting data on something else. To help in gaining conceptual clarity, it would be wise for would-be researchers to think about Internet studies as falling into one of three general categories. First, there are studies of the *Internet*. These are investigations that are primarily focused on the medium or technology itself. Second, there are studies of *experiences* that occur in, on, or through the Internet. These are social and cultural studies that investigate what people do with the technology. Third, there are studies of *people* by use of the Internet. These are studies in which the Internet is a means to an end, a mechanism for soliciting, contacting, or otherwise gaining access to people.

While it is conceivable that researchers may craft studies that fall into any one of these general categories, for all practical purposes, ethnographic research on the Internet is a necessary mixture of all three. It is research on people, about experiences, by use of communication technologies. One can hardly imagine ethnographic research on the Internet that does not entail some degree of each of these things. The point is not that we ought to separate these three aspects of human experience on the Internet, but rather, we should be attentive to which of these is our primary analytical focus so as to appropriately frame our methods and analysis. This distinction is critical. Failure to acknowledge these distinctions can easily result in circumstances where researchers collect data on something that is quite different from what was intended. Even more, there are pragmatic implications: if a researcher seeks to focus on experiences that occur on the Internet, then there are some questions, problems, and information that will prove to be largely irrelevant. Conversely, if a researcher seeks to focus on people by use of the Internet, those very same questions, problems, and information may be extremely relevant. Failure to grasp these distinctions can prove quite crippling because they are not only essential to effective data collection and analysis but also necessary for the more difficult tasks of communicating to others what we have studied and what was learned from those studies.

For example, because I have sought to understand experiences that occur in online chat and cybersex, I have not concerned myself with the "real" people "behind" these online encounters. Given the frame of my analysis, it did not matter who these people "really" were. The "actual" age, gender, race, or any of the other usual demographic variables that are so routinely included in surveys and interviews were unimportant for the purposes of my research—all that mattered was what these informants told me about their experiences. Furthermore, asking such questions potentially threatened participants' anonymity and, therefore, posed a risk to the integrity of my research. So, I never asked. Framing Internet studies in this way is perfectly legitimate. But, the critical thing is that I remain aware of how I have framed my studies and, thus, conscious of what that information can and cannot tell me; what I can and cannot say about it.

Even if researchers are aware of how they have framed their Internet studies that does not mean that others will recognize it. In fact, it is here where it is especially easy to become frustrated, confused, or misled, often by well-intended comments, suggestions, and reviews. On many occasions, I have been told things like, "Your analysis is artificial because you do not know anything about the people you have studied." Comments like these are frustrating because I never intended to say anything about those people in the first place; my studies were not designed to glean that kind of information; it is a kind of review that fails to distinguish between the people who use the Internet and the more general experiences that occur by use of the Internet. One person went so far as to say that I could not legitimately study text cybersex unless I included in my analysis how participants manage the actual orgasms that occur in these experiences, apparently implying that a physical orgasm is necessary for the experience to be considered "sex." I have received many of these kinds of silly comments and reviews and will not belabor it further—my point is that these remarks reflect the kind of misunderstanding that I am driving at: a failure to acknowledge that studying people by use of the Internet is not necessarily the same as studying experiences that occur by use of the Internet; studies of the Internet as a medium and technology are not necessarily the same as studies of the social and cultural significance of the Internet. Because the Internet is a mediated context for communication and social interaction, we not only can make these distinctions, I argue that we must in order to adequately frame our research, clearly see what we are studying, and maximize what we can learn from those inquiries.

Online Chat, Text Cybersex, and Televideo Cybersex:
The Methods and Data of Three Related Studies

The data cited throughout this book were collected as part of three related studies that began in 1994 and concluded in 2002. The first study focused on online chat, the second on text cybersex, and the third on televideo cybersex. These three studies were designed as part of a single analytical scheme, divided into separate methodological projects. Like an onion, I sought to understand personhood in online chat and cybersex as a layered experience, to which each of these three studies represented a distinct analytical peel. The study of online chat was all about the fluidity of selfhood made possible by anonymity and the disembodied nature of the medium. Having analytically connected disembodiment with self-fluidity, the text cybersex study sought to explore how participants evoke a body in these erotic encounters and how these textual semiotic bodies reveal relationships between self, body, and situated social interaction. The televideo cybersex study further explored these relationships in situations where the body appears as an image to be seen.

Although analytically connected, these three studies were conducted as separate methodological projects. In each successive project, I refined analysis, honed methods, and experimented with data collection in several different contexts by use of a variety of methodologies. However, for practical reasons, I ended up favoring some contexts and methods of data collection over others; some places and methods simply proved more useful for the purposes of my research.

For both the chat and text cybersex studies, a large commercial online service was chosen as the primary context for research. This service is one of the largest and most successful commercial Internet service providers. This particular service was selected due to its popularity for providing a user-friendly means of interacting with others in what is (and was more so in the past) a sometimes daunting computer-mediated world of cyberspace. Some data was collected on IRC, which also features vibrant experiences of online chat and cybersex, but IRC was (and to a significant extent remains) arcane and intimidating to the less computer literate. Web-based chat was also considered, but it took a while for the bandwidth and client software to develop. Thus, one commercial service proved to be most useful primarily because it was user-friendly, allowing a larger and more diverse number of people to participate.

Between 1997 and 2002, there were several false starts and unsuccessful attempts at the televideo cybersex study. I had conceptualized and crafted much of the televideo cybersex study by 1997 but was unable to effectively collect data until several years later. These problems were directly related to the difficulties of finding an appropriate context for research. Until very recently, CUSeeMe was the most popular of the few softwares that allow users to connect and chat in televideo. CUSeeMe, however, requires that users connect either directly (assuming they know the IP number of the participant with whom they wish to connect) or through "reflector sites" that are typically owned and operated by individuals or institutions. These reflector sites are not always easy to find, and once found, often turn out to be empty. Most of the popular reflector sites for CUSeeMe—especially for televideo cybersex—are "members only." For these reasons, it proved to be extremely difficult to collect data on televideo cybersex, although several attempts were made. Finally, with the development of new client software and commercial televideo chat services, I was able to complete this study. The majority of that data came from one popular televideo chat service provider whose user-friendly software makes the experience intuitive enough for almost anyone to effectively navigate.

Several different methods of data collection were used, or at least experimented with. Observation and interview methodologies proved most useful. In each of the three studies, substantial time was spent exploring and generally "hanging out" in online chat and cybersex environments. I made no attempt to systemize my observations. Although I occasionally recorded what I observed for use in analysis and discussion, for the most part, my observation and participation in online chat and cybersex was used to gain personal familiarity and understanding of these experiences. It is possible for researchers to accurately portray experiences they have no direct familiarity with, relying entirely on informants' accounts alone, but I have found that there is no substitute for firsthand experience if one wants to fully understand the character of human experiences and convey those subjective meanings to others.

Open-ended qualitative interviews turned out to be the most effective method for data collection, and these interviews were the source of nearly all the data I have reported. I simply spent time in online chat and cybersex environments, often communicating with participants by use of a screen name such as "Research," "Wants2Know," or "CollectingData." I would explain to participants that I was doing a study and ask for their participation. Participants would be invited to contact me—usually by private messages—by which I would conduct a real-time online interview. In each of the three studies, I only used data from informants who agreed to participate in the study.

All participants were briefed on the nature of the study, told they could with-draw from the interview at any time, guaranteed anonymity, and encouraged to ask questions. Participants who requested copies of the study were pro-vided them (once compiled into a coherent manuscript), and their feedback on my analysis was also encouraged.

The unstructured nonstandardized interview strategy was used (Denzin 1989; Lincoln and Guba 1985). This interview style does not use pre-established questions. Instead, interview questions emerge from the re-sponses of participants. This method of data collection relies on a conversa-tional style that is exceptional for allowing the discussion to flow from a participant's perspective (although sometimes the researcher refocuses the direction of the conversation). This unstructured and nonstandardized style also allows for tremendous flexibility and a free-flowing format that adds comfort and familiarity to the interview process. Most importantly, unstruc-tured and nonstandardized interviews approximate the informal conversa-tional structure of most online chat and cybersex environments. This approach is also quite useful for dealing with the sometimes delicate and po-tentially embarrassing issues that are sometimes contained in the experience of cybersex.

A total of fifty-nine interviews were conducted for the study of online chat.[1] The text cybersex study included sixty-two interviews. Thirty-one in-terviews were conducted for the televideo cybersex study. Of these three studies, televideo cybersex proved the most challenging, and it was the only study that involved some demographic assessment of participants (all par-ticipants identified themselves or could be recognized as adults. Twenty-four participants were male and seven were female. Nineteen identified them-selves as heterosexual, six as homosexual, and six as bisexual. No additional demographic information was collected). Televideo cybersex participants are not particularly accessible for research, and therefore, large samples are dif-ficult to obtain. Even in adult televideo chat environments, not all partici-pants are interested in sex, and (obviously) those that are would often prefer to be doing "something else" other than answering probing questions from a researcher. It took over four months to generate thirty-one interviews on televideo cybersex, and I spent a lot of time just waiting for someone willing to participate in the study.

Methodological Advantages and Disadvantages

In my studies, I have learned that doing ethnographic research on online chat and cybersex has methodological advantages and disadvantages, some of which are directly related to the practical nuts and bolts of data collection, while others have to do with more subtle features of the medium. The most significant practical advantage of ethnographic research on the Internet is that, generally speaking, data can be collected from almost anywhere at almost any time. Problems of access are greatly simplified by the inherently dislocated nature of online communication, interaction, and forms of community. Ethnographic research on the Internet does not necessarily require the immediate presence of the researcher in the same physical, geographic, and even temporal space of the people and experiences that he or she is studying. So long as I had access to an Internet connection, I could conduct interviews or otherwise collect data. The advantages are apparent: for the most part, I was able to gather data from my home or office; morning, noon, or night; any day of the week; for any length of time that was convenient for me.

While certainly a distinct advantage, researchers do need to be careful. The convenience of access requires special attention to differences that may or may not influence who and what they are studying. At the very least, researchers need to be aware that different kinds of people are likely to be connected to the Internet at different days of the week and times of the day. In my own research—especially on cybersex—there were distinct qualitative differences between participants who logged on in the morning and early afternoon as opposed to those who logged on at night; there were distinct differences between those who logged on during the weekday and those who connected on the weekends.[2] The key difference boils down to a distinction between those who are employed during normal business hours and those who are unemployed, retired, work non-routine hours, work in the home, or attend school full-time. These kinds of differences may introduce significant bias and may require careful control, depending on the nature and subject of the study.

Another distinct advantage of ethnographic research on the Internet is that the majority of data does not require transcription. As anyone who has experience with interview methodologies will say, transcribing recorded verbal interviews can be extremely time consuming, labor intensive, and tedious. Since most communication on the Internet is in the form of written text, enormous time is saved by the convenient fact that participants transcribe their words for you. This advantage, however, comes at a cost; it introduces new problems. First, there are problems of brevity. Because in most cases, participants must type their responses to interview questions, often these responses are far too brief to be of much ethnographic use. When a participant responds to a question in spoken words, he or she can say a lot in a very short period of time—even if, ironically, they don't have much to say. It takes substantially more time and effort to write or type a reply, and while this does reduce some of the redundancy that is common in spoken interviews, this also tends to strip responses of much of their qualitative richness. Therefore, it is often necessary to enrich responses by probing participant—following up statements with additional related questions in order to get a more complete response for the purposes of analysis.

Hyper-self awareness is another potential problem related to the fact that participants have to type their responses to interview questions. That is, in all likelihood, participants are much more aware of what they are saying and how it reflects onto their self when they have to write or type responses to questions, rather than speak them. While it is true that in speaking we can hear ourselves talk, when we write, our words are much more objectified because we can see what we have written, and furthermore, those words are more easily edited (which, incidentally, is the primary reason why writing is such a valuable learning tool). The situation produces a kind of auto-Hawthorne effect, where in writing or typing the self can much more easily observe itself, pass judgments on what has been written, and edit those responses accordingly. The problem is that responses may be, at least potentially, less spontaneous, more self-reflexive, constrained, conventional, or more consciously crafted to be a positive reflection of the person answering the questions. This is more or less of a problem depending on the subject the researcher is studying and whether or not the researcher assumes (I believe naively) that spontaneous, verbal, off-the-cuff statements are more indicative of an informant's "real" attitudes, beliefs, or opinions. Either way, in many contexts for ethnographic research on the Internet, this kind of hyper-self awareness is nullified by the dislocated and anonymous nature of the medium. If informants are anonymous and not co-present with the interviewer, they may be much less inhibited in ways that would introduce potential bias in face-to-face interview methods. I have come to the conclusion that the realities of interview methodologies on the Internet are a combination of both of these things; participants are extremely self-aware yet also less inhibited in their expressions.

Finally, there is a problem of time and money. Ironically, although typing responses often leads to problems of brevity, even so, researchers must understand that it still takes much longer for participants to respond to questions. In real-time online interviews, one must often wait patiently for what seems a long time for a participant to reply to a question. Furthermore, it can be difficult for respondents to reply to long and complex questions, which are usually better handled through e-mail rather than a real-time interview, or at least broken down into something a little more manageable. In some cases, time spent answering questions may cost participants money. This is less of a problem today, but I know in my early studies both participants and I were paying by the hour for our Internet connection, often outrageously. Either way, whether paying by the hour or not, time spent answering a researcher's question is time spent away from the purposes and motivations for which these participants are using the Internet. This often requires limiting questions, or not asking all participants all questions.

While ethnographic data does not require formal transcription, that does not mean the data does not need to be transposed. The "written-speech" of online chat and cybersex is significantly different from formal text; it is expressed by use of different conventions, grammar, and syntax. I have described these agreed-upon maxims of communication as a kind of "idioculture" that is an integral element of emergent online associations. However, these idiocultural constructs may necessitate that researchers translate data into a form meaningful for reporting. After all, an exact replication of data collected from an online interview or chat room would often either mislead or confuse readers. The common use of the ellipsis, for example, is an agreed upon method for online chat and cybersex participants to indicate long and thoughtful pauses. Unless the researcher omits these ellipses in reporting the data they have collected, readers would be misled into thinking they represent the omission of a portion of what the informant had said (which is how ellipses are typically used in standard grammar). Perhaps more importantly, there is a general idioculture to online chat and cybersex, but specific online groups will form their own unique idiocultural constructs. That is, while certain basic emoticons and acronyms are pervasive (just about anywhere you will find the use of emoticons and acronyms like :) and LOL), various online communities will develop their own distinct paralinguistic forms that are unique to their motives and purposes for communication. I believe that these constructs are an intimate part of how people maintain symbolic group boundaries on the Internet and are key mechanisms used to identify "insiders," but they are often utterly meaningless to those who are not members of their associations. These latter kinds of truly *idio*cultural constructs almost always require that researchers translate the data into a form meaningful for analysis and reporting.

There are various ways that researchers can translate data into a form meaningful for reporting. One can cite data exactly as it appears in its original, adding notes in brackets to explain meaning when necessary. However, I have found this approach to be both distracting and unnecessary. In my work, I simply replace these idiocultural constructs with the appropriate words and grammar necessary to convey the same meaning in the conventions of formal printed text (except in cases where the emoticon or acronym is already widely known and understood). I even go so far as to correct participant's misspellings, typos, and grammar (except in cases where improper grammar is used stylistically). In reporting my data, I do not indicate that I have made these changes. There is nothing about this that I find innately problematic, and changing participants' words in this way does not misrepresent the data. After all, when we conduct studies where we transcribe recorded verbal interviews into printed text, it is necessary to add punctuation and grammar to what informants have told us. Doing so does not misrepresent the data; it merely transfers the meaning of what informants have said to the conventions of a text medium. Such translations are necessary in verbal interviews; so too are they necessary for much data collected in real-time online interviews—we must translate data from its original form of written speech to one of formal text.[3]

Finally, anonymity can also become a practical problem for Internet research. In many online environments, anonymity is highly valued and protected by participants. This is not true of all online associations, but in my studies—especially of cybersex—anonymity was an essential part of the experiences that I sought to understand. Therefore, I found it mandatory to respect this anonymity, which meant that I avoided asking questions that revealed identifying information. These problems can be overcome to whatever extent that researchers become personally known and trusted by participants. But this too will vary depending on what the researcher is studying. In my studies of online chat and cybersex, some participants were routinely present and I came to know them on a more personal level, but most participants would come and go; there was a small number of "regulars" and a majority whose participation was quite ephemeral. Thus, in these conditions, it was impossible to develop trust among the majority of participants. If I were to rely only on information provided by those who came to know and trust me enough to forego their anonymity and disclose personal information, my data would have been seriously skewed and misrepresentative. For these reasons, I found it necessary to simply respect the anonymity of all participants.

Connected to anonymity is a related problem of truth. It has long been suggested to me that the anonymity of online chat and cybersex creates problems with the trustworthiness of participants' claims. That is, there are problems in assessing whether or not people are who they claim to be, and if what they claim is truthful. Frankly, I find this criticism not just silly, but just plain stupid. If this criticism were valid then we would have to ask the same question of all kinds of empirical research, not the least of which are survey methods. Just because someone is given a survey (either in person or sent by snail mail), answers questions with a pen or pencil, then returns the survey to the researcher does not mean they are going to be any more or less truthful than if you e-mail the survey or ask them via real-time communication technologies (telephone, Internet, or otherwise). Survey methodologies contain just as much opportunity for participants to misrepresent themselves as any method of data collection that might be used in Internet studies. Although various theoretical traditions will approach empirical data (or "accounts") differently, for the most part, the majority of empirical research necessitates that researchers suspend at least some degree of doubt concerning the truthfulness of participants' claims. In most cases, failure to do so incapacitates research. Research on the Internet is no different, nor is it any better or worse for procuring reliable and valid data.

NOTES

1. The study I published in 1997 with Mark Douglass in *The Information Society* was a manuscript that I submitted in April of 1996 and was accepted by the editor later that year. That early framing of this study was based on preliminary interviews with nineteen online chat participants. The full study includes fifty-nine interviews.

2. Since I did not ask identifying information or collect demographic data on the participants in my studies, I cannot give specifics on these differences. It is reasonable to assume that different kinds of people will connect at different times of the day and days of the week, and this seemed apparent in my observations and interviews, but I am unable to specify the precise differences. All I suggest is that researchers be aware of this potential source of bias.

3. Obvious exceptions are studies that involve a direct analysis and investigation of these idiocultural constructs and paralinguistic means of constructing and sustaining this blend of writing and talking. As a general rule, however, if a researcher is studying something that is not directly connected to these agreed-upon conventions of written speech, then it is not only appropriate but, in most cases, desirable to transpose original data into the conventions of formal text.

❖ REFERENCES

Abate, T. 1998. "Virtual Voyeurs Webcams Can Open Offices, Bedrooms to World Wide Watchers." *Rocky Mountain News.* June 22, 2B.

Altheide, D. and R. Snow. 1991. *Media Worlds in the Postjournalism Era.* New York, NY: Aldine.

Anderson, W. 1997. *The Future of the Self: Inventing the Postmodern Person.* New York, NY: Penguin.

Associated Press. 1996. "Virtual Affair, A Real Divorce." Newsday. February 2, A05.

Aycock, A., and N. Buchignani. 1995. "The E-Mail Murders: Reflections on 'Dead' Letters." *Cybersociety: Computer-Mediated Communication and Community* (ed. by S. Jones). Pages 184–231. Thousand Oaks, CA: Sage.

Baron, N. 1984. "Computer-Mediated Communication as a Force in Language Change." *Visible Language.* 18(2): 118–141.

Baudrillard, J. 1981. *Simulacra and Simulations* (trans. P. Foss, P. Patton and P. Beitchman). New York, NY: Semiotext.

Bauman, Z. 1991. *Modernity and Ambivalence.* Cornell University Press.

Baumeister, R. 1989. *Masochism and the Self.* Hillsdale, NJ: Erlbaum.

———. 1991. *Escaping the Self: Alcoholism, Spirituality, Masochism, and Other Flights From the Burden of Selfhood.* New York, NY: Basic.

Baym, N. 1995. "The Emergence of Community in Computer-Mediated Communication." *Cybersociety: Computer-Mediated Communication and Community* (ed. by S. Jones). Pages 138–163. Thousand Oaks, CA: Sage.

Bechar-Israeli, H. 1995. "From Bonehead to cLoNehEAd: Nicknames, Play and Identity On Internet Relay Chat," *Journal of Computer-Mediated Communication,* 1 (2), URL: http://shum.huji.ac.il/jcmc/vol1/issue2/vol1no2.html.

Becker, H. 1963. "Becoming a Marijuana User." *Outsiders: Studies in the Sociology of Deviance.* New York, NY: Free Press.

———. 1967. "History, Culture, and Subjective Experience." *Journal of Health and Social Behavior.* 8: 163–176.

Bell, D., and R. Holliday. 2000. "Naked as Nature Intended." *Body and Society.* 6 (3–4): 127–140).

Benedikt, M. 1994. *Cyberspace: First Steps.* Cambridge: MIT Press

Berger, P., and T. Luckmann. 1966. *The Social Construction of Reality: A Treatise in the Sociology of Knowledge.* New York, NY: Doubleday-Anchor.

Brissett, D., and C. Edgley. 1990. *Life as Theater: A Dramaturgical Sourcebook.* New York, NY: Aldine de Gruyter.

Bryant, C. 1982. *Social Deviancy and Social Proscription: The Social Context of Carnal Behavior.* New York, NY: Human Sciences Press.

Cahill, S. 2001. *Inside Social Life: Readings in Sociological Psychology and Microsociology.* Los Angeles, CA: Roxbury.

Cahill, S., W. Distler, C. Lachowetz, A. Meaney, R. Tarallo, and T. Willard. 1985. "Meanwhile Backstage: Public Bathrooms and the Interaction Order." *Urban Life.* 14: 33–58.

Carse, J. 1986. *Finite and Infinite Games.* New York, NY: MacMillan.

Chayko. M. 1993. "What is Real in the Age of Virtual Reality? 'Reframing' Frame Analysis for a Technological World." *Symbolic Interaction.* 16(2): 171–181.

Cheseboro, J. and D. Bonsall. 1989. *Computer-Mediated Communication: Human Relationships in a Computerized World.* Tuscaloosa, AL: University of Alabama Press.

Cohen, S. and L. Taylor. 1992. *Escape Attempts: The Theory and Practice of Resistance to Everyday Life* (second edition). New York, NY: Routledge.

Cooley, C. 1902 [1964]. *Human Nature and the Social Order.* New York, NY: Scribner's.

Cooper, A., D. Delmonico, and R. Burg. 2000. "Cybersex Users, Abusers, and Compulsives: New Findings and Implications." *Cybersex: The Dark Side of the Force* (ed. by A. Cooper). Pages 5–29. Philadelphia, PA: Brunner-Rutledge.

Cooper, A., Putnam, D., L. Planchon, and S. Boies. 1999. "Online Sexual Compulsivity: Getting Tangled in the Net." *Sexual Addiction and Compulsivity.* 6 (2): 79–104.

Davis, M. 1983. *Smut: Erotic Reality/Obscene Ideology.* Chicago, IL: University of Chicago Press.

Denzin, N. 1989. *The Research Act: Theoretical Introduction to Sociological Research Methods.* Englewood Cliffs, NJ: Prentice Hall.

Dery, M. 1992. "Sex Machine, Machine Sex: Mechano-Eroticism and Robo-Copulation." *Mondo 2000: A User's Guide to the New Edge* (ed. by R. Rucker, R. Sirius, and M. Queen). New York, NY: HyperPerennial.

Dewey, J. 1916. *Democracy and Education: An Introduction to the Philosophy of Education.* New York, NY: Macmillian.

Douglas, J. 1977. *The Nude Beach.* Beverly Hills, CA: Sage.

Douglas, M. 1970. *Purity and Danger: An Analysis of Concepts of Pollution and Taboo.* Baltimore, MD: Penguin.

Durkheim, E. 1898 [1974]. "Individual and Collective Representations." *Sociology and Philosophy* (trans. by D.F. Pocock). Pages 1–34. New York, NY: Free Press.

———. 1912 [1965]. *The Elementary Forms of Religious Life.* New York, NY: Free Press.

Eco, U. 1986. *Travels in Hyperreality.* Orlando: Harcourt, Brace, Jovanovich.

Edelsward, L.M. 1991. "We Are More Open When We Are Naked." *Ethnos.* 56 (3–4): 189–199.

Edgley, C. and D. Brissett. 1999. *A Nation of Meddlers.* Boulder, CO: Westview.

Elmer-Dewitt, P. 1994. "Bards on the Internet." *Time Magazine.* July 4, 66–67.

Featherstone, M., and R. Burrows. 1995. *Cyberspace, Cyberbodies, Cyberpunk: Cultures of Technological Embodiment.* Thousand Oaks, CA: Sage.

Festinger, L. 1957. *A Theory of Cognitive Dissonance.* Stanford, CA: Stanford University Press.

Fine, G. 1987. *With the Boys: Little League Baseball and Preadolescent Culture.* Chicago, IL: University of Chicago Press.

Flowers, A. 1998. *The Fantasy Factory: An Insider's View of the Phone Sex Industry.* Philadelphia, PA: University of Pennsylvania Press.

Forsyth, C. 1992. "Parade Strippers: A Note on Being Naked in Public." *Deviant Behavior: An Interdisciplinary Journal.* 13: 391–403.

Foucault, M. 1978. *History of Sexuality: An Introduction* (volume I). New York, NY: Vintage.

———. 1979. *Discipline and Punish: The Birth of the Prison.* New York, NY: Penguin.

Frank, A. 1991. *At the Will of the Body: Reflections on Illness.* Boston, MA: Houghton Mifflin.

Gadow, S. 1982. "Body and Self: A Dialectic." *The Humanity of the Ill: Phenomenological Perspectives* (ed. by V. Kestenbaum). Pages 86–100. Knoxville, TN: University of Tennessee Press.

Geertz, C. 1965. *New Views of the Nature of Man.* Chicago, IL: Chicago Press.

Gergen, K. 1991. *The Saturated Self: Dilemmas of Identity in Contemporary Life.* Basic Books.

———. 1994. *Realities and Relationships: Soundings in Social Construction.* Cambridge, MA: Harvard University Press.

———. 1999. *An Invitation to Social Construction.* Thousand Oaks, CA: Sage.

Gibson, W. 1984. *Neuromancer.* New York, NY: Ace.

Glaser, M. 1998. "Cybertainment: For Those Who Like to Watch, Home Webcam Provides Picture of Private Lives." *Los Angeles Times.* January 8, 41.

Glassner, B. 1990. "Fit for Postmodern Selfhood." *Symbolic Interaction and Cultural Studies* (ed. by H. Becker and M. McCall). Pages 215–243. Chicago: University of Chicago Press.

Goffman, E. 1959. *The Presentation of Self in Everyday Life.* Garden City, NY: Doubleday Anchor.

———. 1961. *Asylums: Essays on the Social Situation of Mental Patients and Other Inmates.* New York, NY: Free Press.

———. 1963a. *Stigma: Notes on the Management of Spoiled Identity.* Englewood Cliffs, NJ: Prentice Hall Touchstone.

———. 1963b. *Behavior in Public Places: Notes on the Social Organization of Gatherings.* New York, NY: Free Press.

———. 1968a. *Asylums: Essays on the Social Situation of Mental Patients and Other Inmates.* New York, NY: Anchor.

———. 1971. *Relations in Public: Microstudies in Public Order.* New York, NY: Basic Books.

———. 1974. Frame Analysis: An Essay of the Organization of Experience. New York, NY: Harper and Row.

Gracyk, T. 1991. "Pornography as Representation: Aesthetic Considerations." *Pornography: Private Right or Public Menace?* (ed. by R. Baird and S. Rosenblum). Pages 117–137. Buffalo, NY: Prometheus.

Gubar, S. 1989. "Representing Pornography." *For Adult Users Only: The Dimensions of Violent Pornography* (ed. S. Gubar and J. Hoff). Pages 47–67. Bloomington: Indiana University Press.

Gunn, A. 1994. "After Hours." *PC Magazine.* April: 441–443.

Hayles, K. 1993. "Virtual Bodies and Flickering Signifiers." *October,* 66–91.

Healy, D. 1996. "Cyberspace and Place: The Internet as Middle Landscape on the Electronic Frontier." *Internet Culture* (ed. by D. Porter). Pages 55–68. New York, NY: Routledge.

Heim, M. 1991. "The Erotic Ontology of Cyberspace." *Cyberspace: First Steps* (ed. by M. Benedikt). Pages 59–80. Cambridge: MIT Press.

Henslin, J., and M. Biggs. 1971. "The Sociology of the Vaginal Examination." *Studies in the Sociology of Sex* (ed. by James Henslin). New York, NY: Appleton, Century, Crofts.

Hertzler, J. 1965. *A Sociology of Language*. New York, NY: McGraw-Hill.

Hochschild, A. 1983. *The Managed Heart*. Berkley, CA: University of California Press.

Holstein, J., and J. Gubrium. 1994. "Grounding the Postmodern Self." *Sociological Quarterly*. 35 (4): 685–703.

———. 2000. *The Self We Live By: Narrative Identity in a Postmodern World*. New York, NY: Oxford University Press.

Horton, D., and R. Wohl. 1956. "Mass Communications and Para-Social Interaction: Observations on Intimacy at a Distance." *Psychiatry*. 19: 215–229.

James, W. 1892 [1961]. *Psychology: The Briefer Course*. New York, NY: Harper and Brothers.

Jones, R. 1994. The Ethics of Research in Cyberspace. *Internet Research*. 4(3):30–35.

Jones, S. 1995. *Cybersociety: Computer-Mediated Communication and Community*. Thousand Oaks: Sage.

———. 1998. *Cybersociety 2.0: Revisiting Computer-Mediated Communication and Community*. Thousand Oaks, CA: Sage.

Kanaley, R. 1998. "The World is Getting Stranger: They're Baring Their Souls, and Much More, on the Web." *Philadelphia Inquirer*. March 26.

Katz, J. 1988. *Seductions of Crime: Moral and Sensual Attractions in Doing Evil*. Basic Books.

Katz, M. 1998. "Internet Surfing Pulls Users Away From Family, Activities; Experts Debate Normalcy of Excessive Computer Use." *The Washington Times*. October 20, A2.

Kiesler, S., J. Siegel, and T. McGuire. 1984. "Social Psychological Aspects of Computer-Mediated Communication." *American Psychologist*. 39(10): 1123–1134.

King, S. 1996. "Researching Internet Communities: Proposed Ethical Guidelines for the Reporting of Results." *The Information Society: An International Journal*. 12(2): 119–127.

Kling, R. 1996. *Computerization and Controversy: Value Conflicts and Social Choices*. San Diego, CA: Academic Press.

Krielkamp, T. 1976. "Erving Goffman: The Constraints of Socialization." *The Corrosion of Self* (Chapter Five). New York, NY: New York University Press.

Kristeva, J. 1987. *In the Beginning Was Love: Psychoanalysis and Faith*. New York, NY: Columbia University Press.

Krivel, P. 1998. "Peek-a-Boo: Is The Internet Turning us Into a Planet of Voyeurs and Exhibitionists?" *Toronto Star*. July 23.

Kupfer, J. 1983. *Experience as Art: Aesthetics in Everyday Life*. Albany, NY: State University of New York Press.

Lane, F. 2000. *Obscene Profits: The Entrepreneurs of Pornography in the Cyber Age*. New York, NY: Routledge.

Langner, L. 1991. *The Importance of Wearing Clothes*. Los Angeles, CA: Elysium.

Laurel, B. 1993. *Computers as Theater*. Reading, PA: Addison-Wesley.

Leiblum, S. 1997. "Sex and the Net: Clinical Implications." *Journal of Sex Education and Therapy*. 22 (1): 21–27.

Lifton, R. 1993. *The Protean Self: Human Resilience in an Age of Fragmentation*. New York, NY: Basic Books.

Lincoln, Y., and E. Guba. 1985. *Naturalistic Inquiry*. Newbury Park, CA: Sage.

Loseke, D. 1987. "Lived Realities and the Construction of Social Problems: The Case of Wife Abuse." *Symbolic Interaction*. 10: 229–243.

MacKinnon, R. 1995. "Searching for the Leviathan in Usenet." *Cybersociety: Computer-Mediated Communication and Community* (ed. by S. Jones). Pages 112–137. Thousand Oaks, CA: Sage.

Mantovani, G. 1996. *New Communication Environments: From Everyday to Virtual*. Bristol, PA: Taylor and Francis.

Markham, A. 1998. *Life Online: Researching Real Experience in Virtual Space*. Lanham, MD: Rowman and Littlefield.

Marx, G. 1994. "New Telecommunications Technologies and Emergent Norms." *Self, Collective Behavior and Society: Essays in Honor of Ralph Turner*. JAI Press.

Mason-Schrock, D. 1996. "Transsexuals' Narrative Construction of the 'True Self.'" *Social Psychology Quarterly*. 59 (3): 176–192.

May, R. 1975. *The Courage to Create*. New York, NY: W. W. Norton.

McCormick, N., and J. Leonard. 1996. "Gender and Sexuality in the Cyberspace Frontier." *Women and Therapy*. 19 (4): 109–119.

McLuhan, M. 1964. *Understanding Media: Extensions of Man*. Cambridge, MA: MIT Press.

McLuhan, M., and Q. Fiore. 1967. *The Medium is the Massage: An Inventory of Effects*. New York, NY: Bantam.

McRae, S. 1996. "Flesh Made Word: Sex, Text and the Virtual Body." *Internet Culture* (ed. by D. Porter). Pages 73–86. New York, NY: Routledge.

Mead, G. 1934. *Mind, Self, and Society* (ed. by C. Morris). Chicago, IL: University of Chicago Press.

Merleau-Ponty, M. 1962. *Phenomenology of Perception* (trans. by C. Smith). London: Routledge and Kegan Paul.

Meyrowitz, J. 1985. *No Sense of Place: The Impact of Electronic Media on Social Behavior*. New York, NY: Oxford University Press.

Moore, T. 1998. *The Soul of Sex: Cultivating Life as an Act of Love*. New York, NY: Harper Collins

———. 1997. "Sex (American Style)." *Mother Jones*. September/October.

Morris, D. 1991. *The Culture of Pain*. Berkeley, CA: University of California Press.

Myers, D. 1987. "'Anonymity is Part of the Magic': Individual Manipulation of Computer-Mediated Communication Contexts." *Qualitative Sociology*. 10(3): 251–266.

Nagel, T. 1979. "Sexual Perversion." *The Journal of Philosophy*. 66 (1): 5–17.

Nakamura, L. 2000. "'Where Do You Want To Go Today?': Cybernetic Tourism, the Internet, and Transnationality." *Race in Cyberspace* (ed. by B. Kolke, L. Nakamura, and G. Rodman). Pages 15–26. New York, NY: Routledge.

Norman, D. 1993. In *Computers as Theater* (by B. Laurel). Pages xi–xv. Reading, Mass: Addison-Wesley.

Norwich, C. 1987. "Parts Plus." *Photo/Design*. July-August: 51–54.

Nussbaum, M. 1995. "Objectification." *Philosophy and Public Affairs*. 24 (4): 249–291.

Odzer, C. 1997. *Virtual Spaces: Sex and the Cyber Citizen.* New York, NY: Berkley Books.

Ogburn, W. 1964. *On Culture and Social Change.* Chicago, IL: University of Chicago Press.

Penny, S. 1994. "Virtual Reality as the Completion of the Enlightenment Project." *Culture on the Brink: Ideologies of Technology* (ed. G. Bender and T. Druckrey). Seattle: Bay Press.

Porter, D. 1996. *Internet Culture.* New York, NY: Routledge.

Poster, M. 1997. "Cyberdemocracy: Internet and the Public Sphere." *Internet Culture* (ed. by D. Porter). Pages 201–217. New York, NY: Routledge.

Postman, N. 1988. "Social Science as Moral Theology." *Conscientious Objections: Stirring up Trouble About Language, Technology, and Education.* Pages 3–19. New York, NY: Vintage Books.

———. 1992. *Technopoly: The Surrender of Culture to Technology.* New York, NY: Vintage.

Reid, E. 1991. "Electropolis: Communication and Community on Internet Relay Chat." Honors Thesis. University of Melborne, Australia.

———. 1994. *Cultural Formations in Text-Based Virtual Realities.* Master's Thesis. University of Melbourne: Department of English.

Resnick, R. 1992. "The Electronic Meet Market." *Compute.* August, 90–91.

Rheingold, H. 1991. *Virtual Reality.* New York, NY: Touchstone.

Rice, R. 1984. *The New Media: Communication, Research, and Technology.* Beverly Hills, CA: Sage.

———. 1989. "Issues and Concepts in Research on Computer-Mediated Communication Systems." *Communication Yearbook 12* (ed. by J. Anderson). Pages 436–476. Newbury Park, CA: Sage.

Rival, L., D. Slater, and D. Miller. 1998. "Sex and Sociality: Comparative Ethnographies of Sexual Objectification." *Theory, Culture, and Society.* 15 (3–4): 295–321.

Robinson, P., and N. Tamosaitis. 1993. *The Joy of Cybersex: An Underground Guide to Electronic Erotica.* New York, NY: Brady.

Rose, L. 1995. *Netlaw: Your Rights in the Online World.* Berkeley, CA: Osborne McGraw-Hill.

Rucker, R., R. Sirius, and M. Queen. 1993. *Mondo 2000: A User's Guide to the New Edge.* New York, NY: HyperPerennial.

Sanderson, D., and D. Sanderson. 1993. *Smileys.* New York, NY: O'Reilly and Associates.

Sartre, J.P. 1956. *Being and Nothingness.* New York, NY: Philosophical Library.

Schneider, J. and R. Weiss. 2001. *Cybersex Exposed: Simple Fantasy or Obsession?* Center City, MN: Hazelden.

Shapiro, M. and D. McDonald. 1992. "I'm Not a Real Doctor, But I Play One in Virtual Reality: Implications of Virtual Reality for Judgments About Reality." *Journal of Communication.* 42(4): 94–114.

Shilling, C. 1993. *The Body and Social Theory.* London: Sage.

Simmel, G. 1950. *The Sociology of Georg Simmel* (ed. by K. Wolff). New York, NY: Free Press.

———. 1971 [1904]. "Fashion." *Georg Simmel: On Individuality and Social Forms* (ed. by D. Levine). Chicago: University of Chicago Press.

———. 1971 [1911]. "The Adventurer." *Georg Simmel: On Individuality and Social Forms* (ed. by D. Levine). Pages 187–198. Chicago: University of Chicago Press.

Slater, D. 1998. "Trading Sexpics on IRC: Embodiment and Authenticity on the Internet. *Body and Society.* 4 (4): 91–117.

Smith, A., and S. Kleinman. 1989. "Managing Emotions in Medical School: Students' Contacts With the Living and the Dead." *Social Psychology Quarterly.* 52 (1): 56–69.

Smith, D. 1990. *Texts, Facts, and Femininity: Exploring the Relations of the Ruling.* London: Routledge.

Solomon, R. 1974. "Sexual Paradigms." *The Journal of Philosophy.* 71 (11): 336–345.

Spear, P. 1991. "Online Games People Play." *Compute.* November, 96–100.

Sproull, L. and S. Kiesler. 1991. *Connections: New Ways of Working in the Networked Organization.* Cambridge, Mass: MIT Press.

Steinberg, D. 1996. "Inside the Noisy World of Online Chat." *Virtual City.* Winter.

Stone, A. 1994. "Will the Real Body Please Stand Up?: Boundary Stories About Virtual Cultures." *Cyberspace: First Steps* (ed. by M. Benedikt). Pages 81–118. Cambridge: MIT Press.

———. 1995. *The War of Desire and Technology at the Close of the Mechanical Age.* Cambridge: MIT Press.

Strauss, A. 1964. *George Herbert Mead on Social Psychology.* Chicago: University of Chicago Press.

———. 1993. *Continual Permutations of Action.* New York, NY: Aldine De Gruyter.

Symonds, C. 1971. "A Nude Touchy-Feely Group." *The Journal of Sex Research.* 7 (2): 126–133.

Thomas, W. I. 1966. "The Relation of Research to the Social Process." *W.I. Thomas on Social Organization and Social Personality* (ed. by M. Janowitz). Pages 289–305. Chicago: University of Chicago Press.

Tisdale, S. 1994. *Talk Dirty to Me: An Intimate Philosophy of Sex.* New York, NY: Anchor.

Toolan, J., M. Elkins, D. Miller, and P. D'Encarnacao. 1974. "The Significance of Streaking." *Medical Aspects of Human Sexuality.* 8: 152–65

Trachtenberg, S. 1985. *The Postmodern Movement: A Handbook of Contemporary Innovation in the Arts.* Westport: Greenwood Press.

Tribe, L. 1991. "The Constitution in Cyberspace: Law and Liberty Beyond the Electronic Frontier." *The Humanist.* September/October: 15–21.

Turkle, S. 1984. *The Second Self: Computers and the Human Spirit.* New York, NY: Simon and Schuster.

———. 1995. *Life on the Screen: Identity in the Age of the Internet.* New York, NY: Simon and Schuster.

Turner, B. 1991. "Recent Developments in the Theory of the Body." *The Body: Social Process and Cultural Theory* (ed. by M. Featherstone, M. Hepworth, and B. Turner). Pages 1–35. London: Sage.

Turner, V. 1969. *The Ritual Process: Structure and Anti-Structure.* Ithaca, NY: Cornell University Press.

Ullman, J. 1998. "Cybersex (Trial Shows Post-Modern Courtship as Reflected in E-Mail Exchanges)." *Psychology Today.* 31: 28.

Van Gelder, L. 1985. "The Strange Case of the Electronic Lover." *Ms.* October, Pages 94–95.

Walther, J. 1992. "Interpersonal Effects in Computer-Mediated Interaction: A Relational Perspective." *Communication Research.* 19(1): 52–90.

Walther, J. and J. Burgoon. 1992. *Pragmatics of Human Communication: A Study of Interactional Patterns, Pathologies, and Paradoxes.* New York, NY: Norton.

Wardhaugh, R. 1985. *How Conversation Works.* Blackwell

Waskul, D. 2002. "Naked Self: Being a Body in Televideo Cybersex." *Symbolic Interaction.* 25 (2): 199–227.

Waskul, D., and M. Douglass. 1996. "Considering the Electronic Participant: Some Polemical Observations on the Ethics of Online Research." *Information Society: An International Journal.* 12 (2): 129–39.

———. 1997. "Cyberself: The Dynamics of Online Chat." *Information Society: An International Journal.* 13 (4): 375–97.

Waskul, D., M. Douglass, and C. Edgley. 2000. "Cybersex: Outercourse and the Enselfment of the Body." *Symbolic Interaction.* 23 (4): 375–397.

Watts, A. 1966. *The Book: On the Taboo Against Knowing Who You Are.* New York, NY: Random House.

Weeks, L. 1997. "Web Site for Voyeur Eyes." *The Washington Post.* September 20, Pages H01.

Weinberg, M. 1965. "Sexual Modesty, Social Meanings, and the Nudist Camp." *Social Problems.* 12 (3): 311–318.

———. 1966. "Becoming a Nudist." *Psychiatry.* 29 (1): 15–24.

———. 1967. "The Nudist Camp: Way of Life and Social Structure." *Human Organization.* 26 (3): 91–99.

Weinstein, E. and P. Deutschberger. 1963. "Some Dimensions of Altercasting." *Sociometry.* 26: 545–566.

Whittle, D. 1996. *Cyberspace: The Human Dimension.* New York, NY: W. H. Freeman and Company.

Williams, L. 1989. *Hardcore: Power, Pleasure, and the "Frenzy of the Visible."* Los Angeles, CA: University of California Press.

Williams, R. 1976. *Keywords: A Vocabulary of Culture and Society.* New York, NY: Oxford University Press.

Wolf, N. 1990. *The Beauty Myth: How Images of Beauty are Used Against Women.* New York, NY: William Morrow and Company, Inc.

Young, K. 1998. *Caught on the Net: How to Recognize the Signs of Internet Addiction—and a Winning Strategy for Recovery.* New York, NY: John Wiley and Sons.

Zurcher, L. 1977. *The Mutable Self: A Self-Concept for Social Change.* Beverly Hills, CA: Sage.

❖ INDEX

face-to-face, 139–140;
  para-social, 139;
  play, 7–10, 48–50
Compulsivity, cybersex, 17–18n, 93n
Compuserve, 5
Computer-mediated communication,
  as dislocated/disembodied, 121, 138–139
Cons/imposters, 46–47

Conversation:
  as similar to sex, 9;
  pleasures of, 8–9
Cooper, Al, 93n
Cooley, Charles, 112, 126
Coquetry, 8
Creativity, heightened, 33, 36
Cross dressing, virtual, 5
Cues filtered out: 32–38;
  shortcomings of, 33, 37–38
Cultural crack: 57–58, 136–137;
  negotiations of, 59–69, 136–137
Cultural lag, 57–58, 137
CuSeeMe, 147
Cyberself: 19–54 (*also see* Online chat);
  and anonymity, 44–48;
  as true self, 63–64, 66;
  consistency of, 54n;
  defined, 43–44;
  multiplicity/fluidity of, 44–45, 48
Cybersex: 71–130, 137–138;
  and sexuality, 103–106;
  as dislocated/disembodied, 73;
  as experience that contradicts form, 72;
  as magnifying body-self-society relationships, 73–74, 96;
  as transgression, 103–105;
  compulsivity, 17–18n, 93n;
  contrast between text and televideo, 95–96;
  defined as experience, 71–72
Cybersex (televideo): 95–130, 137–138;
  and desirability of body, 110;
  and re-enchantment of body, 110–114;
  and screen names, 119;
  as like pornography, 106–107, 116;

described, 100;
  eroticism of, 113;
  impersonality of, 116–117;
Cybersex (text): 71–94;
  and anonymity, 80–81;
  and breast size, 88, 94n;
  and criminality, 52;
  and descriptions of body, 87–88;
  and multiplicity/fluidity of self, 80–81;
  and screen names, 80–81;
  and subject bodies, 74, 85, 89;
  and virtual body, 84–89;
  as requiring literacy, 79;
  as safe sex, 83–84;
  value of, 82
Cyberplace, 22
Cyberspace: 20–24;
  and social reality, 21–24;
  as less real, 23;
  as something *sui generis*, 22;
  defined, 21
Cyborgs, 92

Data/data collection (*see* Research)
Dating, online, 2
Davis, Murray, 75, 109, 112, 113–114
Deceit, 46–47, 110–111
Dery, Mark, 117
Desire, 112
Dewey, John, 32
Diaries, online, 4–5
Disembodiment:
  and self multiplicity/fluidity, 44–46, 77, 89–90, 92;
  and virtuality, 90–93;
  as liberating, 85–86, 89–90;
  as safe sex, 83–84;
  in online chat; 44–45;
  dramaturgies of, 91–92
Domination/oppression, 127n, 134–135
Doubt/suspicion, in online chat, 46–47
Douglas, Jack, 100–102, 107, 112, 114–115, 122
Douglass, Mark, ix, 19, 51, 143, 153
Drama, and self, 14–15
Dramaturgy, 12, 14–15

James, William, 99
Jesus, seeks loving woman, 2–3

Katz, Jack, 128n

Lane, Frederick, 130n
Langner, Lawrence, 116, 130n
Language:
 and symbolic interaction, 12–13;
 discursive/non-discursive, 14
Laurel, Brenda, 33, 47–48, 78
Le viol, 128–129n
Lifton, Robert, 61
Liminality, and public nudity, 101
List, the, 5
Looking glass, 112, 137
Love, online, 2
Luckmann, Thomas, 55, 60–61, 139

Magritte, Rene, 128–129n
Markham, Annette, 20, 48
Mardi Gras, 100–101
Marriage, 112
Marx, Karl, 19, 53n
Mason-Schrock, Douglas, 98, 99–100
May, Rollo, 36
McLuhan, Marshall, 10–11, 19, 25, 141
Me, the, 114
Mead, George H., 13–14, 81, 97, 98, 127n
Meaning:
 in dramaturgy, 14;
 in symbolic interaction, 12–13
Medium:
 is the message, 19–20;
 meta, 143–144
Men, in televideo cybersex, 118–121
Meyrowitz, Joshua, 20, 25, 84, 85, 124, 140
Miller, Daniel, 104–105
Mind-body distinction, 94n
Monogamy, 112
Moralists, 17–18n, 23
MUD, 17n, 53–54n
Multiphrenia, 62–63

Nagel, Thomas, 112

Names, significance of for self, 38–40;
 screen (*see* Screen names/nicks)
Naked:
 as sacred, 122–124, 130n;
 being, 110, 113–114;
 being seen, 107–108
Naked self, 109–110
Nude:
 beaches, 100–102, 107;
 body politics of, 120–121, 130n
Nudity:
 and embodiment, 109–110;
 and role removal, 109–110;
 as cultural heresy, 122–124;
 as self-reduction, 113–114;
 deviance of, 102–103, 130n;
 eroticism of, 102–103, 107–109, 112;
 public, 100–102;
 made antierotic, 101–102;
 therapeutic value of, 112–113
Nudism, 100–102

Object body (*see* Body)
Obscenity, 5
Odzer, Cleo, 3–4, 10, 11, 87
Ogburn, William, 57–58
Online chat: 2, 19–54;
 acronyms in, 34;
 and anonymity, 44–48;
 and doubt/suspicion, 46–47;
 and emoticons, 34–36;
 and idioculture, 31–38;
 and presentation of self, 40–43;
 and private messages, 31;
 and self-multiplicity/fluidity, 55–69;
 as challenge to unitary self, 56–57;
 as dislocated/disembodied, 51–53, 55;
 as form of interaction 28–31, 50–51;
 multiplicity of simultaneous
 conversations, 28–31;
 playfulness of, 42, 48–50;
 synchronicity of, 29
Online diaries, 4–5
Oppression, 127n, 134–135
Outercourse, 72–74

General Editor: Steve Jones

Digital Formations is the new source for critical, well-written books about digital technologies and modern life. Books in this series will break new ground by emphasizing multiple methodological and theoretical approaches to deeply probe the formation and reformation of lived experience as it is refracted through digital interaction. Each volume in *Digital Formations* will push forward our understanding of the intersections—and corresponding implications—between the digital technologies and everyday life. This series will examine broad issues in realms such as digital culture, electronic commerce, law, politics and governance, gender, the Internet, race, art, health and medicine, and education. The series will emphasize critical studies in the context of emergent and existing digital technologies.

For additional information about this series or for the submission of manuscripts, please contact:

Acquisitions Department
Peter Lang Publishing
275 Seventh Avenue 28th Floor
New York, NY 10001

To order other books in this series, please contact our Customer Service Department:

(800) 770-LANG (within the U.S.)
(212) 647-7706 (outside the U.S.)
(212) 647-7707 FAX

or browse online by series:

WWW.PETERLANGUSA.COM